SELECTED TOPICS IN WIMAX

Edited by **Gianni Pasolini**

Selected Topics in WiMAX
http://dx.doi.org/10.5772/56712
Edited by Gianni Pasolini

Contributors

Tassio Carvalho, Jose Jailton Junior, Mohammed Awadh Ben-Mubarak, Elankayer Sithirasenan, Varun Jeoti, Mohammad Mehdi Gilanian Sadeghi, Mona Shokair

Published by InTech

Janeza Trdine 9, 51000 Rijeka, Croatia

Publishing Process Manager Ana Pantar
Technical Editor InTech DTP team
Cover InTech Design team

Additional hard copies can be obtained from orders@intechopen.com

Selected Topics in WiMAX, Edited by Gianni Pasolini
p. cm.
ISBN 978-953-51-1157-3

For EU product safety concerns:
IN TECH d.o.o., Prolaz Marije Krucifikse Kozulić 3, 51000 Rijeka, Croatia,
info@intechopen.com or visit our website at intechopen.com.

Meet the editor

Gianni Pasolini received the M.Sc. degree in Telecommunications Engineering and the Ph.D. degree in Electronic Engineering and Computer Science from the University of Bologna, Italy, in 1999 and 2003, respectively.
In May 1999 he joined the IEIIT (the Institute for Electronics and for Information and Telecommunications Engineering of the National Research Council). Here he started working on wireless local and personal area networks (WLAN and WPAN), WLANs and WPANs coexistence, WLAN radio resource management, WLANs/UMTS integration and WiMAX. His research interests include also digital signal processing and Intelligent Transportation Systems.
In October 2006 he became a Researcher at the University of Bologna where he has been teaching telecommunications since 2003.

Contents

Preface

In June 2001, operators and equipment vendors in the communications ecosystem founded the nonprofit WiMAX Forum, an industry-led organization aimed at harmonizing broadband wireless access standards, and certifying interoperability among devices from different manufacturers.

Nowadays, about 10 years later, the WiMAX technology is a mature and affordable solution for high-speed IP-based 4G mobile broadband, fully supporting bandwidth-intensive services, such as high-speed Internet access and television, as well as less bandwidth-demanding but more latency-sensitive services, such as voice-over-IP calls.

The deployment of WiMAX networks, which are based on the IEEE 802.16-2004 Air Interface Standard, started in many countries in 2005, as soon as manufacturers released the first equipments. The early WiMAX networks were aimed at providing fixed broadband wireless services in a wide geographical scale and proved to be a really effective solution for the establishment of wireless metropolitan area networks.

In February 2006, the IEEE802.16e-2005 amendment gave a further extension to the potentialities of this technology, introducing a number of features aimed at supporting also users mobility; in this way, the so-called Mobile-WIMAX profile was born. The result was a complete standard family that specified the air interface for both fixed and mobile broadband wireless access, enabling thus the convergence of fixed and mobile networks through a common wide area radio access technology.

Since then, the interest in WiMAX has increased dramatically; nowadays, this technology is at work in more than 100 countries, providing both residential and mobile connectivity in urban areas as well as remote locations, due to the relatively low costs associated with its deployment (in comparison with 3G, HSDPA).

The parallel development of the 4G Long Term Evolution (LTE) wireless access technology, currently deployed in many countries, is not expected to displace WiMAX, which had a first-mover advantage. It is likely, in fact, that the two technologies will be complementary.

As LTE networks begin to roll out, it is extremely likely, for instance, that WiMAX will be used as the wireless backhaul for those networks, leaving to LTE the access provision. Moreover, WiMAX seems preferable for DSL replacement, especially in developing countries and rural regions, where wireline broadband technologies are not available. For these reasons, both the industrial and the scientific communities keep on developing and investigating WiMAX, with the final aim to get the best from this technology.

In this book, a collection of selected papers is presented, which covers several aspects of the WiMAX technology, investigated by specialists working in universities and research centres all over the world.

Chapter 1 is focused on multiuser (MU) multiple input multiple output (MIMO) diversity techniques. It provides a comprehensive overview of various MIMO strategies and a critical discussion on the recent research in multiuser MIMO communications with emphasis on the design of precoders. State-of-the-art in research on MU-MIMO precoding and multi-cell processing is presented. A new precoder design and its application in WiMAX setting is also provided.

In chapter 2 the issue of the high peak-to-average power ratio (PAPR) that affects WiMAX is investigated, with particular reference to the adoption of companding as a PAPR mitigation strategy. Impairments from nonlinear distortion at the transmitter, multipath and noise are considered.

Chapter 3 deals with the adoption of mesh architectures within WiMAX, and provides an overview of the main challenges of the WiMAX mesh mode, with a focus on routing protocols and the effect of quality of service mechanisms on scenarios with mobile clients.

Chapter 4 investigates the handover mechanism, with particular reference to the cell selection scheme. A multi-criteria selection scheme is proposed, based on the experienced carrier-to-interference plus noise ratio, the required bandwidth and the expected congestion delay.

In chapter 5 the issue of coordinated authentication in a heterogeneous network environment is investigated. A Coordinated Robust Authentication (CRA) Mechanism is presented, with the ability to use a single set of credentials with any network.

Finally, Chapter 6 reviews the Multicast and Broadcast re-keying Algorithm (MBRA) of IEEE 802.16e, that ensures the confidentiality of communications. The operational efficiency, scalability and 1-affects-n phenomenon are discussed.

Gianni Pasolini, Ph.D., Researcher.
Department of Electrical, Electronic, and Information Engineering "Guglielmo Marconi".
University of Bologna, Italy

Chapter 1

On MU-MIMO Precoding Techniques for WiMAX

Elsadig Saeid, Varun Jeoti and Brahim B. Samir

Additional information is available at the end of the chapter

http://dx.doi.org/10.5772/56034

1. Introduction

Future wireless communication systems will require reliable and spectrally efficient transmission techniques to support the emerging high-data-rate applications. The design of these systems necessitates the integration of various recent research outcomes of wireless communication disciplines. So far, the most recent tracks of investigations for the design of the spectrally efficient system are multiple antennas, cooperative networking, adaptive modulation and coding, advanced relaying and cross layer design. The original works of Telatar [1], Foschini [2] and the early idea of Winters [3] as well as the contribution of Almouti[4] stress the high potential gain and spectral efficiency by using multiple antenna elements at both ends of the wireless link. This promising additional gain achieved by using Multiple Input Multiple Output (MIMO) technology has rejuvenated the field of wireless communication. Nowadays this technology is integrated into all recent wireless standards such as IEEE 802.11n, IEEE 802.16e and LTE [5, 6]. The main objective of this chapter is to provide a comprehensive overview of various MIMO Techniques and critical discussion on the recent advances in Multi-user MIMO precoding design and pointing to a new era of the precoding application in WiMAX systems.

The chapter opens with the basic preliminaries on MIMO channel characterization and MIMO gains in section 2. This is followed by theoretical overview of precoding for MU-MIMO channel in section 3. Section 4 describes a new precoding method and numerical simulation stressing the importance of the proposed precoding in the WiMAX context. Section 5 concludes the chapter.

2. Preliminaries

2.1. Single user MIMO (SU-MIMO) channel model

In wireless communications, channel modeling and link parameter design are the core problems in designing communication system. To understand the MIMO system, MIMO channel modeling and the related assumptions behind the practical system realization should therefore be first discussed and summarized. Basically, it all begins by the designer identifying the channel type from among the three types of wireless communication channel namely: direct path, frequency selective and frequency-flat channels. Direct path, also called Line-of-Sight (LOS) channel is the simplest model where the channel gain consistes of only the free space path loss plus some complex Additive White Gaussian Noise (AWGN). This simplified model is typically used to design the various microwave communication links such as terrestrial, near space satellite, and deep space communication links. The other two types of wireless channel models are the frequency flat and frequency selective channels which both describes the channel gains due to the complex propagation environment where both the free space path loss, shadowing effects and multipath interference are obvious and the objective of the link model is to account for multipath and Doppler shift effects. In frequency selective channel model, the link gain between each transmit and receive antennas is represented by multiple and different impulse response sequences across the frequency band of operation. This is in contrast to the frequency flat fading which has single constant scalar channel gain across the band. Frequency selectivity has crucial Inter-Symbol Interference (ISI) effect on the high speed wireless communication system transmission. Technically there are three ways to mitigate the negative effect of ISI. Two of which are transmission techniques namely: spread spectrum transmission and Multicarrier modulation transmission while the third is the equalization techniques as a receiver side mitigation method. In a MIMO system, researchers generally make common assumption that the channel-frequency-response is flat between each pair of transmit and receive antennas. Thus, from the system design point of view, the system designer can alleviate the frequency selectivity effect in a wideband system by subdividing the wideband channel into a set of narrow sub-bands as in [7, 8] using Orthogonal Frequency Division Multiplexing (OFDM). Figure 1 shows point-to-point single user MIMO system with N_T transmit antennas and $M_{\mathbf{R}}$ receive antennas. The channel from the multiple transmit antenna to the multiple receiving antenna is described by the gain matrix **H**. With the basic assumption of frequency-flat fading narrow band link between the transmitter and the receiver, **H** will be given by:

$$\mathbf{H} = \begin{bmatrix} h_{11} & \cdots & h_{1,N_T} \\ \vdots & \ddots & \vdots \\ h_{M_R,1} & \cdots & h_{M_R,N_T} \end{bmatrix} \in \mathbf{C}^{M_R \times N_T} \tag{1}$$

where $\mathbf{h}_{i,j}$ denotes the complex channel gain between the jth transmit antenna to the ith receive antenna, while $1 \le j \le N_T$, $1 \le i \le M_{\mathbf{R}}$. We further assume that the channel bandwidth is equal to ω. Thus, the received signal vector $\mathbf{y}[n] \in \mathbf{C}^{M_{\mathbf{R}} \times 1}$ at the time instants $[n]$ can be given by:

$$\mathbf{y}[n] = \mathbf{H}\mathbf{s}[n] + \mathbf{n}[n] \tag{2}$$

where $\mathbf{s}[n] = [\mathbf{s}_1 \cdots \mathbf{s}_{N_T}]^T \in \mathbf{C}^{N_T \times 1}$ denotes the complex transmitted signal vector, and $\mathbf{n}[n] \in \mathbf{C}^{M_{\mathbf{R}} \times 1}$ denotes the AWGN vector which is assumed to have independent complex Gaussian elements with zero mean and variance $\sigma_n^2 \mathbf{I}_{M_{\mathbf{R}}}$ where $\sigma_n^2 = \omega N_o$ and N_o is the noise power spectral density.

The channel matrix **H** is assumed to have independent complex Gaussian random variables with zero mean and unit variance. This statistical distribution is very useful and reasonable assumption to model the effect of richly scattering environment where the angular bins are fully populated paths with sparsely spaced antennas. Justification of these MIMO link assumptions is a very important point for system design. On choosing system operation bandwidth ω, the system designers and researchers always assume that the channel frequency response over the bandwidth of MIMO transmission is flat, as practically, it is very hard to implement equalizers for mitigating ISI across all the multiple antennas in a MIMO system. It is to be noted that the twin requirements of broadband transmission to support the high rate applications and narrow band transmission to facilitate the use of simple equalizers to mitigate ISI can be met by utilizing OFDM based physical layer transmission. In physically rich scattering environment (e.g. typical urban area signal propagation) with proper antenna array spacing, the common assumption is that the elements of the Channel matrix are independent identically distributed (i.i.d.). Though, beyond the scope of this study, in practice, if there is any kind of spatial correlation, this will reduce the degrees of freedom of the MIMO channel and consequently this would then result in a decrease in the MIMO channel capacity gain [3, 9]. The assumption of i.i.d channel is partially realizable by correctly separating the multiple element antennas. In some deployment scenarios, where there are not enough scatterers in the propagation environment, the i.i.d assumption is not practical to statistically model the MIMO fading correlation channel. A part of the last decade research was focused on the study of this kind of channel correlation effects [10]. In general, fading correlation between the elements of MIMO channel matrix **H** can be separated into two independent components, namely: transmit correlation and receive correlation [11]. Accordingly, the MIMO channel model **H** can be described by:

$$\mathbf{H} = \mathbf{R}_r^{1/2} \mathbf{H}_w \mathbf{R}_t^{1/2} \tag{3}$$

where $\mathbf{H}_w$ is the channel matrix whose elements are i.i.d, and $\mathbf{R}_r^{1/2}$ and $\mathbf{R}_t^{1/2}$ are the receive and transmit correlation matrix respectively.

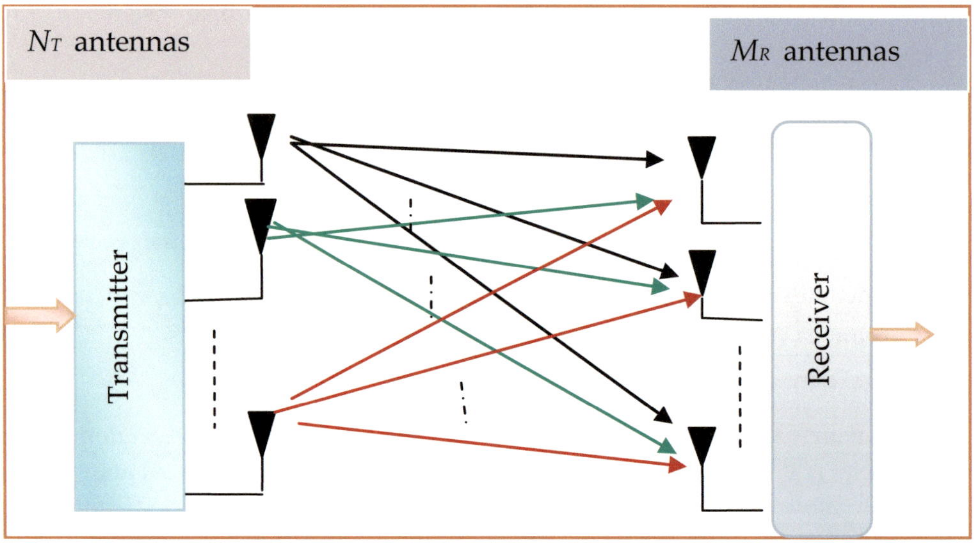

Figure 1. Block Diagram of single User MIMO System

2.1.1. MIMO channel gain

The key to the performance gain in MIMO systems lies in the additional degree-of-freedom provided by the spatial domain and associated with multiple antennas. These additional degree-of-freedom can be exploited and utilized in the same way as the frequency and time resources have been used in the classical Single Input Single Output (SISO) systems. The initial promise of an increase in capacity and spectral efficiency of MIMO systems ignited by the work of Telatar [1] and Foschini [2] has now been validated where by adding more antennas to the transmitter and receiver, the capacity of the system has been shown to increase linearly with the N_T or $M_{\mathbf{R}}$,which is minimum, i.e the min(N_T, $M_{\mathbf{R}}$) [12]. This capacity can be extracted by making use of three transmission techniques, namely: spatial multiplexing, spatial diversity, and beam-forming.

From classical communication and information theory, channel characteristics play a crucial rule in the system design, in that both transmitter and receiver design are highly dependent on it [13, 14]. In MIMO system, the knowledge of the Channel State Information (CSI) is an important factor in system design. CSIT and CSIR refer to the CSI at the transmitter and receiver respectively. Basically, in the state-of-art communication system design, there is a common assumption that the receiver has perfect CSI. With this fair assumption, all MIMO performance gains are exploited. Further improvement in the performance is dependent on the availability and quality of CSI at the transmitter [8, 15]. Accordingly, the accessibility and utilization of CSI at the transmitter is one of the most important criteria of MIMO research classification in the last decade. Next sub-sections gives a brief overview of the most critical processing techniques and types of gains that we can extract from the single user point-to-point MIMO link in both open-loop systems (CSI is available at the receiver) and closed-loop systems (CSI is available at both transmitter and receiver).

2.1.2. Open-loop single user MIMO (SU-MIMO) transmission

When there is no CSI at the transmitter, this is called open-loop MIMO configuration. There are two types of performance gains that can be extracted - multiplexing gain and diversity gain [16]. Multiplexing gain is the increase in the transmission rate at no cost of power consumption. This type of gain is achieved through the use of multiple antennas at both transmitter and receiver. In a single user MIMO system with spatial multiplexing gain configuration, different data streams can be transmitted from the different transmit antennas simultaneously. At the receiver, both linear and nonlinear decoders are used to decode the transmitted data vector. Spatial multiplexing gain is very sensitive to long-deep channel fades. Thus, in such communication environment, the designer can solve this problem by resolving to system design that can extract MIMO diversity gain with the help of time or frequency domain.

Diversity gain is defined as the redundancy in the received signal [17]. It affects the probability distribution of received signal power favorably. In single user MIMO system, diversity gain can be extracted when replicas of information signals are received through independent fading channels. It increases the probability of successful transmission which, in turn increases the communication link reliability. In the single user MIMO system, there are two types of diversity methods that are popular, namely: transmit diversity and receive diversity.

Receive diversity is applied on a sub-category of MIMO system where there is only one transmit antenna and $M_{\mathbf{R}}$ receive antennas, also called Single Input Multiple Output (SIMO). In this case the MIMO channel $\mathbf{H}$ is reduced to the vector of the form:

$$\mathbf{H} = \mathbf{h} = [h_1 h_2 \cdots h_{M_R}] \tag{4}$$

with $\mathbf{s}$ denoting the transmitted signal with unit variance, the received signal $\mathbf{y} \in \mathbf{C}^{M_{\mathbf{R}} \times 1}$ can be expressed as:

$$\mathbf{y} = \mathbf{h}s + \mathbf{n} \tag{5}$$

The received signal vector from all receiving antennas is combined using one of the many combining techniques like Selection Combining (SC), Maximal Ratio Combining (MRC) or Equal Gain Combining (EGC) to enhance the received Signal to Noise Ratio (SNR) [18]. The most notable drawback of these diversity techniques is that most of the computational burden is on the receiver which may lead to high power consumption on the receiver unit.

On the other hand, MIMO transmit-diversity gain can be extracted by using what is called Space Time Codes (STC) or Space Frequency Code (SFC) [12, 19, 20]. Unlike receive diversity, transmit diversity requires simple linear receive processing to decode the received signal. STC and SFC are almost similar in many aspects except that one of them uses the time domain while the other uses frequency domain. Space-time codes are further classified into Space-Time Block Codes (STBC) and Space-Time Trellis Codes (STTC) families. In general, STTC families achieve better performance than STBC families at the cost of extra computational load. A well known example and starting point for understanding the STBC transmit diversity techniques is the basic method of Alamouti code [4] which has diversity gain of the order of $2M_{\mathbf{R}}$. However, the main limitation of the basic Alamouti method is that it works only for two transmit antennas. However, latest advances in MIMO diversity techniques extends this method to the case of MIMO channel with more than two transmit antennas through what is known today as Orthogonal Space Time Block Codes (OSTBC) [21].

2.1.2.1. Channel capacity of open-loop single user MIMO system

Without the CSI at the transmitter, the MIMO channel capacity is defined and obtained in [1, 22, 23]. Specifically, for the time-invariant communication channel, the capacity is defined as the maximum mutual information between the MIMO channel input and the channel output and is given by:

$$C = \omega \log_2 \left\| \mathbf{I} + \frac{1}{\sigma_n^2} \mathbf{H}\mathbf{R}_s\mathbf{H}^* \right\| \text{ bits/s} \tag{6}$$

where ω is bandwidth in Hz and $\mathbf{R}_\mathbf{s}$ is the covariance matrix of the transmitted signal and $P_T = tr(\mathbf{R}_\mathbf{s})$ is the total power-constraint. So, for the single user MIMO channel with a Gaussian random matrix with i.i.d elements, the channel capacity will be maximized by distributing the total transmit power over all transmit antennas equally. Thus, in this uniform power allocation scenario, the input covariance matrix $\mathbf{R}_\mathbf{s}$ must be selected such that:

$$\mathbf{R}_s = \frac{P_T}{N_T} \mathbf{I}_{N_T} \tag{7}$$

With power constraint inequality of the form $tr(\mathbf{R}_\mathbf{s}) \le P_T$, where:$P_T$ is the total transmitting power, the substitution of the power constraint in the average capacity formula of equation (6) yields:

$$C = \omega \log_2 \left\| \mathbf{I} + \frac{1}{\sigma_n^2} \frac{p_T}{N_T} \mathbf{H}\mathbf{H}^* \right\| \text{ bit/s} \tag{8}$$

[18], and in the case of SIMO configuration (one transmit antenna and $M_{\mathbf{R}}$ receive antennas) the channel capacity reduces to:

$$C_{\text{SIMO}} = \omega \log_2 \left(1 + \frac{P_T}{\sigma_n^2} \|\mathbf{h}\|_F^2\right) \text{ bit/s} \tag{9}$$

Andconsequently, for the case of MISO configuration (N_T transmit antenna and one receive antennas) the channel capacity reduces to:

$$C_{SIMO} = \omega \log_2 (1 + \frac{1}{\sigma_n^2} \frac{p_T}{N_T} \|\mathbf{h}\|_F^2) \text{ bit/s} \tag{10}$$

Conversely, for the time varying communication channel, the capacity in equation (8) becomes random or ergodic [7] and is defined by:

$$C_{\text{ergodic}} = E\{\omega \log_2 \left\| \mathbf{I} + \frac{1}{\sigma_n^2} \frac{p_T}{N_T} \mathbf{HH}^* \right\| \} \text{ bit/s} \tag{11}$$

Unlike the capacity gains defined in equations (8-10) which can be extracted by spatial multiplexing or diversity, the system capacity in equation (11) is unidentified and it has no significant practical meaning. Thus, in such cases, the system designer can use some kind of system outage metric for the performance evaluation. The quantity called the outage capacity can be defined by the probability that the channel mutual information is less than some constant C:

$$pro_{\text{out}} = prob\{\mathbf{H} : \omega \log_2 \left\| \mathbf{I} + \frac{1}{\sigma_n^2} \frac{p}{N_T} \mathbf{HH}^* \right\| < C\} \tag{12}$$

2.1.3. Closed-loop single user MIMO transmission

When the CSI is available at both transmitter and receiver, all kinds of MIMO gains (diversity, spatial multiplexing and beam forming) can be extracted and optimized. In practice, CSI can be acquired at the transmitter either through feedback channels in Frequency Division Duplex (FDD) systems or just taking the dual transpose of the received channel in the case of time-invariant Time Division Duplex (TDD) systems[24]. To extract the maximum spatial multiplexing gain, transmission optimization should be done by what is called channel precoder and decoder [25, 26]. For single user MIMO channel, firstly the precoder is designed, multiplied with the user's data, and launched through N_T transmit antennas at the transmitter site. At the receiver, the received signal from the $M_{\mathbf{R}}$ receive antennas is processed by the optimized linear decoder. The general form of the precoded received signal is written as:

$$\mathbf{y} = \mathbf{HFs} + \mathbf{n} \tag{13}$$

where **F** is the transmit precoding matrix. Different constraints and conditions are used to design the single user MIMO precoding matrix. Generalized method of joint optimum precoder and decoder for single user MIMO system based on Minimum Mean Square Error (MMSE) approach is proposed in [15]. In this method, minimum mean square error performance criteria is used. As the name suggests, the framework is general and leads to flexible solution for performance criterias such as minimum BER and maximum information rate. The main drawbacks of this method are its high computational complexity and the restrictions on the number of antennas. In addition, there are many other simple and linear methods of precoding such as zero forcing, Singular Value Decomposition (SVD) [6, 8] or code book based techniques [27, 28]. Although these methods are simple, they have quite acceptable performance. On the other hand, the spatial diversity gain can also be optimized by precoding when some kind of CSI is available at the transmitter. The precoding across the space-time block code in [19] or transmit antenna selection method in [29] are two other notable closed-loop spatial diversity gain optimization techniques.

2.1.3.1. Channel capacity of the closed-loop single user MIMO

Consider the general capacity formula for MIMO system given in [2]:

$$C = \{\log_2 \left\| \mathbf{I} + \frac{1}{\sigma_n^2} \mathbf{H}\mathbf{R}_s\mathbf{H}^* \right\| \} \text{ bit/s} \tag{14}$$

This capacity depends on the channel realization **H** and the input covariance matrix $\mathbf{R_s}$. Taking into account the availability of the CSI at the transmitter, there exists for any practical channel realization, an optimum choice of the input covariance matrix $\mathbf{R_s}$such that the channel capacity is maximized subject to the transmit power constraint [7]. This capacity is calculated from the following optimization:

$$\begin{aligned} & C = \max \sum_i \omega \log_2 (1 + \frac{\lambda_i P_i}{\sigma_n^2}) \\ & \text{subject to:} \\ & \sum_i P_i \le P_T \end{aligned} \tag{15}$$

where λ_i is the ith eigenvalue of the single user MIMO covariance matrix ($\mathbf{HH}^*$) and $\mathbf{P}_i$ is the transmit power on the ith channel. For the purpose of generalization, we assume that the rank of the covariance matrix ($\mathbf{HH}^*$) is (r), so $i = 1 \cdots r$. The solution of the optimization problem given in equation (15) [7] shows that the maximum capacity is achieved by what is called the water-filling in space solution which is given by:

$$\frac{P_i}{P_T} = \begin{cases} \frac{1}{\gamma_o} - \frac{1}{\gamma_i} & \gamma_i \geq \gamma_o \\ 0 & \gamma_i < \gamma_o \end{cases}$$
$$\text{given that:}$$
$$\gamma_i = \frac{\lambda_i P}{\sigma_n^2}, \quad \gamma_o \, is \text{ some cut-off SNR} \tag{16}$$

In short this meants that, technically we have to allocate more power to the strong eigen modes and less power to the weak ones. It is also clear that this capacity is proportional to the min(N_T, $M_{\mathbf{R}}$).

2.2. Multiuser MIMO channel

Unlike the simple SU-MIMO channel, Multi-user MIMO (MU-MIMO) channel is a union of a set of SU-MIMO channels. In MU-MIMO system configuration, there are two main communication links - the downlink channel (One-to-many transmission link) which is also known as MU-MIMO Broadcast Channel (MU-MIMO-BC) and the uplink channel (Many-to-One Transmission link) which is also known as MU-MIMO Multiple Access (MU-MIMO-MAC) channel. In addition to the conventional MIMO channel gains, in MU-MIMO we can make use of the multi-user diversity gain to send simultaneously to a group of users or receive data from multiple users at the same time and frequency. As depicted in figure 2, MU-MIMO system configuration can be described as follows: Central node/base station equipped with N_T transmit antennas transmitting simultaneously to B number of users in the downlink MU-MIMO-BC channel, where k^{th} user is equipped with M_k receive antennas, $k=1, \cdots, B$. In the reverse uplink MU-MIMO-MAC the base station receiving data from the multiple users simultaneously.

Regardless of its implementation complexity, it is generally known that the minimum mean-square-error with successive interference cancelation (MMSE-SIC) multi-user detector is the best optimum receiver structure for the MU-MIMO-MAC channel [30]. To simultaneously transmit to multiple users in the downlink MU-MIMO-BC, Costa's Dirty-Paper Coding (DPC) or precoding is needed [31] to mitigate the Multi-User Interference (MUI). Both linear and nonlinear precoding transmission techniques have been heavily researched in the last decade with much preference given to the linear precoding methods owing to their simplicity [32-37]. In the downlink MU-MIMO-BC channel at some k^{th} user, the received signal is given by:

$$\mathbf{y}_k = \mathbf{H}_k \mathbf{F}_k \mathbf{s}_k + \mathbf{H}_k \sum_{\substack{i=1 \\ i \neq k}}^{B} \mathbf{F}_i \mathbf{s}_i + \mathbf{n}_k \tag{17}$$

where $\mathbf{H}_k \in \mathbf{C}^{M_k \times N_T}$ is the channel from the base station to the kth user. $\mathbf{F}_k \in \mathbf{C}^{N_T \times M_k}$ is the kth user precoding matrix, while $\mathbf{s}_k \in \mathbf{C}^{M_k \times 1}$ is the kth user transmitted data vector and $\mathbf{n}_k$ is the received additive white noise vector at the kth user antenna front end.

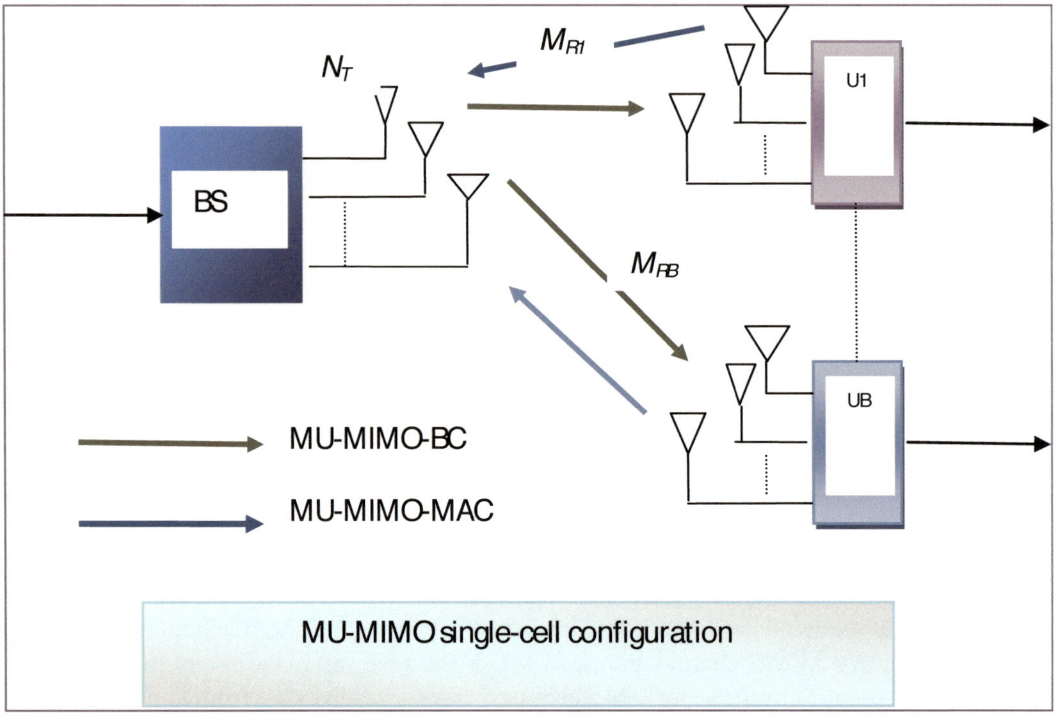

Figure 2. Block Diagram of MU-MIMO System Configuration

Before returning to the general question like how to design the precoder, what is the best Precoder, and what we can gain by incorporating multi-user mode in the WiMAX standards, we summarize the different MIMO system configurations in figure 3. Basically, there are two main modes, SU-MIMO and MU-MIMO. Essentially MU-MIMO is a closed-loop transmission system which means that the channel-state information is required at the transmitter for any transmission. There are two modes of operation for SU-MIMO configurations – closed-loop where the CSI is required at the transmitter and open-loop where the CSI is not required to be used at the transmitter. Also the diagram indicates at each end the type of MIMO gain that can be extracted by each mode of operation and specific configuration.

2.3. On MIMO receiver

MIMO receiver design is also one of the hot areas of wireless communication research and system development in the last decade. Many receiving techniques have been reported to decode these kinds of vector transmissions. For the linear transmission techniques, the decoders design complexities range from simple linear methods like Zero-Forcing (ZF) and Minimum Mean-Square-Error (MMSE) receivers to complex sphereical sub-optimal decoding and optimal Maximum Likelihood Detections (MLD) [8, 10].

2.3.1. Zero-forcing MIMO receiver

Zero-Forcing (ZF) decoder is a simple linear transformation of the received signal to remove the inter-channel interference by multiplying the received signal vector by the inverse of the channel matrix [38]. In fact, if perfect CSI is available at the receiver, the zero-forcing estimate of the transmitted symbol vector can be written as

$$\overline{\mathbf{y}} = \mathbf{G}(\mathbf{H}\mathbf{s} + \mathbf{n}) = \mathbf{s} + \mathbf{G}\mathbf{n} \tag{18}$$

where the decoder is calculated from $\mathbf{G} = (\mathbf{H}^*\mathbf{H})^{-1}\mathbf{H}^*$, which is also known as the pseudo inverse of the MIMO channel matrix. In ZF, the complexity reduction comes at the expense of noise enhancement which results in some performance losses compared to other MIMO receiving methods.

2.3.2. Minimum mean-square-error MIMO receiver

Unlike ZF receiver which completely force the interference to zero, the MIMO MMSE receiver tries to balance between interference mitigation and noise enhancement [39]. Thus, at low SNR values the MMSE outperforms the ZF receiver. In the MMSE MIMO receiver the decoding factor **G** is designed to maximize the expectation criteria of the form:

$$E\{[\mathbf{G}\mathbf{y} - \mathbf{s}][\mathbf{G}\mathbf{y} - \mathbf{s}]^*\} \tag{19}$$

By analytically solving this MMSE criterion for MIMO channel, the factor **G** is found to be:

$$\mathbf{G} = (\mathbf{H}^*\mathbf{H} + \sigma_n^2\mathbf{I})^{-1}\mathbf{H}^* \tag{20}$$

With successive Interference Cancelation (SIC), additional nonlinear steps are added to the original ZF and MMSE equalizers. The resulting versions are ZF-SIC and MMSE-SIC decoding methods. In short, in SIC, the data layer symbols are decoded and subtracted successively from the next received data symbol starting with the highest SINR received signal at each decoding stage. The main drawback of this kind of receive structure is however, the error propagation.

2.3.3. Maximum likelihood MIMO receiver

The Maximum Likelihood (ML) decoder is an optimum receiver that achieves the best BER performance among all other decoding techniques. In ML, the decoder searches for the input vector **s** that minimizes the ML criteria of the form:

$$\left\| \mathbf{y} - \mathbf{H}\mathbf{s} \right\|_F^2 \tag{21}$$

where $\| . \|_F^2$ denotes the matrix/or vector Frobenius norm. The complexity of this decoder increases exponentially as the number of transmit and receive antennas increases. In spite of its good BER performance, ML decoding is however not used in any practical system.

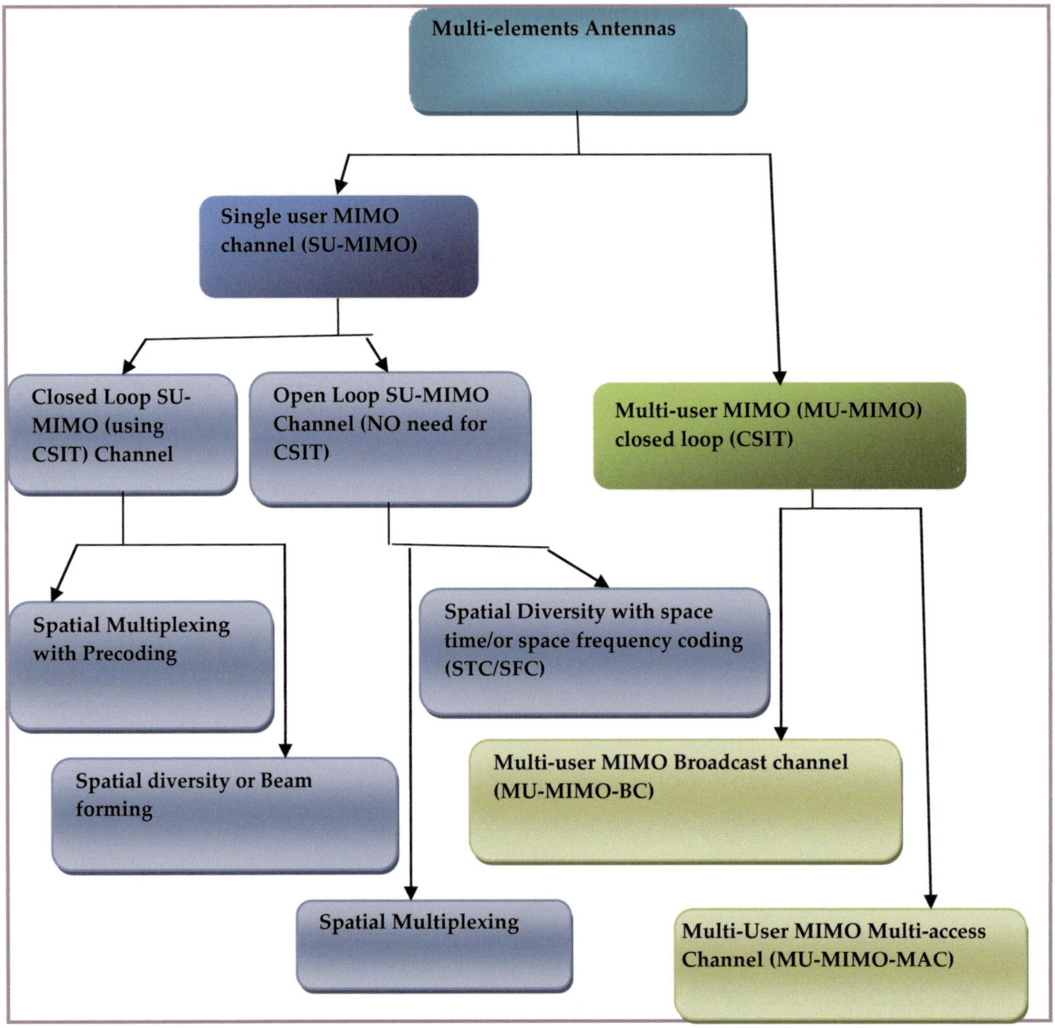

Figure 3. Different forms of Multi-element Antennas Channel Configuration

2.3.4. Sphere decoding MIMO receiver

Sphere Decoding (SD) families are the new decoding techniques that aim to reduce the computational complexity of the ML decoding technique. In the sphere decoder, the received signal is compared to the closest lattice point, since each codeword is represented by a lattice point. The number of lattice points scanned in a sphere decoder depends on the initial radius of the sphere. The correctness of the codeword is in turn dependent on the SNR of the system. The search in Sphere decoding is restricted by drawing a circle around the received signal in a way to encompass a small number of lattice points. This entails a search within sub-set of the codes-words in the constellation and allows only those code-words to be checked. All code-words outside the sphere are not taken into consideration for the decoding operation [8].

2.3.5. MIMO in the current WiMAX standard

MIMO techniques have been incorporated in all recent wireless standards including IEEE 802.16e, IEEE 802.16m, IEEE 802.11n, and the Long-Term Evaluation (LTE). The WiMAX profile IEEE 802.16e defines three different single user open loop transmission schemes in both uplink and downlink channel summarized as below:

- Scheme defined as matrix *A* which describes spatial multiplexing mode of operation for two different symbol streams through two different antennas.
- Scheme defined as matrix *B* which describes the spatial diversity mode of operation for two different symbol streams through two different antennas with the basic Alamouti Space-Time Block Code (STBC) [4].
- Scheme defined as matrix **C** which combines the respective advantages of diversity and spatial multiplexing modes of operation for two different symbol streams through two different antennas. More details of these schemes are given in [40-42].

In addition to the basic MIMO techniques supported by the IEEE 802.16e, the profile of IEEE 801.16m also supports several advanced MIMO techniques including more complex configurations of SU-MIMO and MU-MIMO (spatial Multiplexing and beam-forming) as well as a number of advanced transmit diversity [43, 44]. The profile also defines multi-mode capability to adapt between SU-MIMO and MU-MIMO in a predefined and flexible manner. Furthermore, flexible receiver decoding mode selection is also supported. Unitary precoding or beam-forming with code-book is also defined for both SU-MIMO and MU-MIMO configurations [45]. Cade-book based MU-MIMO precoding techniques found to be effective for the FDD mode of operation because of the great amount of reduction on feedback channel provided, while they are ineffective in the TDD mode of operation [24]. In the next section, we will introduce non-unitary MU-MIMO precoding method to the area of WiMAX. It can be shown that the non-unitary precoding like our proposed method will be applicable and suitable to the TDD mode of operation as accurate CSI is available at the transmitter for the precoding design. In the next section, we will review the most recent researched precoding methods and extend them by proposing our new method.

3. Linear precoding for MU-MIMO system

Keeping in mind the computational complexity of the nonlinear DPC precoding methods, the research community, as we mentioned before, gives more preference to the investigations of computationally simple linear precoding techniques. Many design metrics and conditions are used to develop these linear precoding methods that are hard to deeply survey in one chapter. Generally, one can divide the MU-MIMO linear precoding methods in the literature into two categories - methods that formulate the design objective function for both the precoder and decoder independently such as the methods in [30, 46, 47] and methods that jointly design both the precoding and decoding matrices at the transmitter site (also called iterative method), such as the work in [20, 47-52]. In spite of the good performance joint precoding design obtains relative to the independent formulation methods, the downlink channel overload and complexity are the main drawbacks of this kind of design. One more possible classification is to distinguish between formulations that lead to a closed-form solution expressions such as the works in [53-55] versus those that lead to iterative solutions such as the works in [47, 50, 56, 57]. For comparison, formulations leading to iterative solutions tends to have higher computational complexity than closed form solution methods that are linear. Among the state-of-the-art methods in recent research works, the precoding method originally proposed by Mirette M. Sadek in [34] and based on Per-User Signal to Leakage plus Noise Ratio- Generalized Eigenvalue Decomposition (SLNR-GEVD) and its computationally stable extended version that appeared in [58] which is based on Per-User Signal to Leakage plus Noise Ratio-Generalized Singular Value Decomposition (SLNR-GSVD) are the best in performance. In the next section, we will review these state-of-the-art linear precoding method that seek to maximize Per-user Signal to Leakage plus Noise Ratio (SLNR), which will then be followed by a detailed derivation of our proposed precoding method which is based on maximizing Per-Antenna Signal to Leakage plus Noise Ratio (PA-SLNR) followed by simulation results under WiMAX Physical layer assuming the TDD mode of operation.

3.1. Precoding by signal-to- leakage-plus-noise ratio maximization based on GEVD computation

This precoding method is based on maximizing Signal-to-Leakage-plus-Noise Ratio (SLNR) proposed by [34, 59]. In MU-MIMO-BC, recall the system description of section 2.2 and figure 2. The received signal at the kth user is given by:

$$\mathbf{y}_k = \mathbf{H}_k \mathbf{F}_k \mathbf{s}_k + \mathbf{H}_k \sum_{\substack{j=1 \\ j \neq k}}^{B} \mathbf{F}_j \mathbf{s}_j + \mathbf{n}_k \tag{22}$$

In this received signal expression, the first term represents the desired signal to the kth user, while the second term is the Multi-User Interference (MUI) from the other users to the kth user and the third term is the additive white Gaussian noise at the kth user antenna front end. In the per-user SLNR precoding method, various variables used in the method are depicted in figure 4. The objective function is formulated such that the desired signal component to the kth user, $\| \mathbf{H}_k \mathbf{F}_k \|_F^2$ is maximized with respect to both the signal leaked from the kth user to all other users in the system $\sum_{\substack{j=1 \\ j\neq k}}^{B} \mathbf{H}_j\mathbf{F}_k$ plus the noise power at the kth user front end which is given by $M_{\mathbf{R}_k}\sigma_n^2$. Thus the SLNR objective function for the kth user can be written as:

$$SLNR_k = \frac{\left\|\mathbf{H}_k\mathbf{F}_k\right\|_F^2}{M_{R_k}\sigma_n^2 + \sum_{\substack{j=1 \\ j\neq k}}^{B}\left\|\mathbf{H}_j\mathbf{F}_k\right\|_F^2} \tag{23}$$

By defining the kth user auxiliary interference domain matrix $\widetilde{\mathbf{H}}_k$ as:

$$\tilde{\mathbf{H}}_k = [\mathbf{H}_1 \cdots \mathbf{H}_{k-1}\mathbf{H}_{k+1} \cdots \mathbf{H}_B]^T \tag{24}$$

the precoding matrix $\mathbf{F}_k$ obtained by per-user SLNR maximization is defined as:

$$\mathbf{F}_k = \arg\max_{\mathbf{F}_k} \frac{(\mathbf{F}_k^*\mathbf{H}_k^*\mathbf{H}_k\mathbf{F}_k)}{\mathbf{F}_k^*(\tilde{\mathbf{H}}_k^*\tilde{\mathbf{H}}_k + M_{R_k}\sigma_n^2\mathbf{I}_{N_T})\mathbf{F}_k} \tag{25}$$

Closed form solution is developed to solve this fractional rational mathematical optimization problem by making use of the Generalized Eigenvalue Decomposition (GEVD) technique. Unlike the conventional precoding formulations, this method relaxes the constraints on the number of transmitting antennas and has better BER performance.

3.2. Precoding by signal-to- leakage-plus-noise ratio maximization based on GSVD computation

One important point of observation in GEVD computation is that it is sensitive to the matrix singularity. Thus, the resulting computation accuracy is low. To resolve the singularity problem in the computation of the multi-user precoding matrices from the Per-user SLNR performance criteria, the work in [58, 60, 61] proposes a Generalized Singular Valve Decomposition (GSVD) and QR- Decomposition (QRD) based methods that both overcome the singularity problem and produce numerically better results. Basically, both the GSVD algorithm and QRD based methods optimize the same Per-user SLNR objective function of equation (25) but they handle the singularity problem in the covariance matrix differently. Thus, the final computation result is accurate and the calculated precoder is more efficient in inter-user interference mitigation. The reason behind the singularity problem in the leakage power plus the noise covariance matrix is that at the high SNR value, the power dominates across the matrix which reduces the degree of freedom.

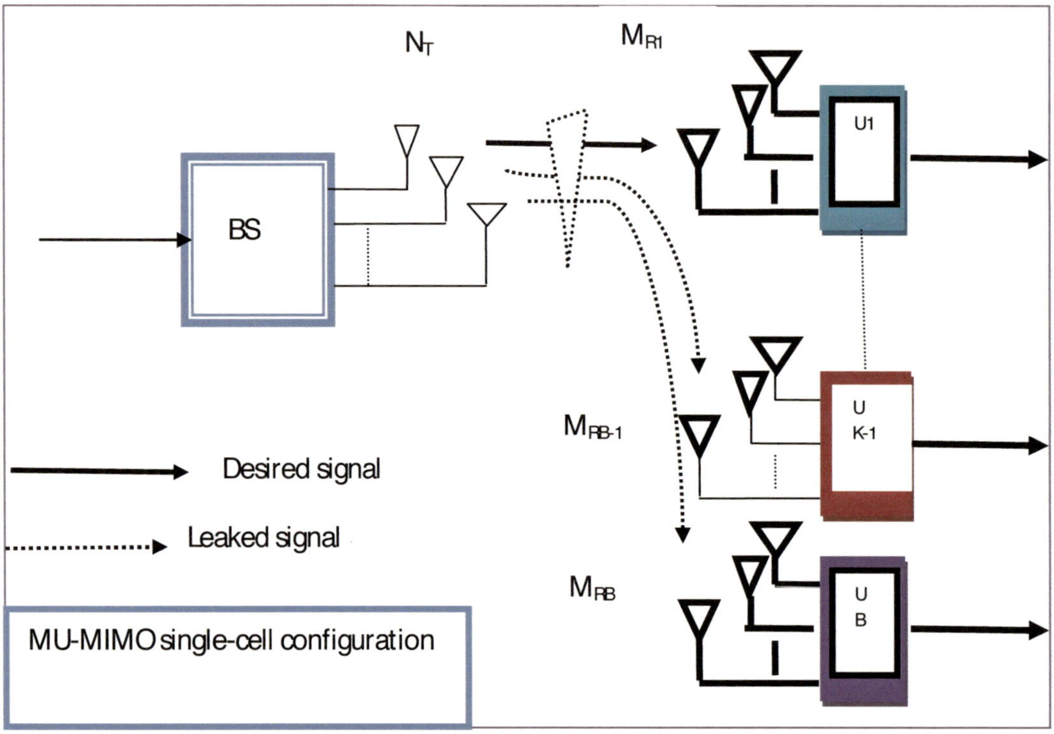

Figure 4. The Definition of the Desired Signal and the Leaked Signal

Two algorithms to solve the objective function of equation (25) are given in [58, 60, 61]. Because of the singularity problem, both algorithms avoid matrix inversion to overcome the computational instability. The developed algorithms makes use of the GSVD analysis. Although there are several methods of GSVD formulations in the literature [62-64], the work in [60] and [58] makes use of the least restrictive form of GSVD algorithm due to Paige and Saunders [62] which is now summarized as follows:

Theorem 1: Paige and Saunders GSVD:

Consider any two matrices of the form: $A_b \in \mathbf{C}^{D\times C}$ and $A_w \in \mathbf{C}^{D\times N}$. The GSVD is given by:

$$Y^T A_b^T Q = [\Sigma_b, \mathbf{0}] \text{ and } Z^T A_w^T Q = [\Sigma_w, \mathbf{0}] \tag{26}$$

$$\text{where } \Sigma_b = \begin{bmatrix} \mathbf{I}_b & & \\ & \mathbf{D}_b & \\ & & \mathbf{0}_b \end{bmatrix} \text{ and } \Sigma_w = \begin{bmatrix} \mathbf{I}_w & & \\ & \mathbf{D}_w & \\ & & \mathbf{0}_w \end{bmatrix} \tag{27}$$

Y and Z are orthogonal matrices and Q is the non-singular eigenvectors matrix. Thus, we can write:

$$A_b^T = Y[\Sigma_b,\ \mathbf{0}]Q^{-1} \text{ and } A_w^T = Z[\Sigma_w,\ \mathbf{0}]Q^{-1} \tag{28}$$

where: $\mathbf{D}_b = \mathrm{diag}(\alpha_1\alpha_2 \cdots \alpha_r)$ and $\mathbf{D}_w = \mathrm{diag}(\beta_1\beta_2 \cdots \beta_r)$, and $\alpha_i + \beta_i = 1,\ i = 1,\ \cdots,\ r$.

The authors in [58, 60] made use of this theorem and developed an efficient precoding algorithm to calculate the precoding matrices for multiple users which is summarized in algorithm 1 as follows

Algorithm 1: The GSVD based Per-user SLNR Precoding algorithm [58, 60]

Assume that the combined channel matrix for MU-MIMO broadcast channel of B number of users is given by $\mathbf{H}_{\mathrm{com}}$ and the input noise power is given by σ_k^2.

1. Input: $\mathbf{H}_{com} = [\mathbf{H}_1;\ \cdots;\ \mathbf{H}_B]$, and σ_k^2

2. Output: The algorithm computes the precoding matrices for Busers.

   a Set $\mathbf{\Psi} = [\mathbf{H};\ \sqrt{M_k}\sigma_k\mathbf{I}_{N_T}] \in \mathbf{C}^{(BM_k+N_T)\times N_T}$

   b Compute the reduced QRD of $\mathbf{\Psi}$ i.e $\mathbf{\Omega}^{\mathbf{H}}\mathbf{\Psi} = \mathbf{R}$ where $\mathbf{\Omega} \in \mathbf{C}^{(BM_k+N_T)\times N_T}$orthonormal columns and $\mathbf{R} \in \mathbf{C}^{N_T\times N_T}$ is upper triangular.

3. For $k = 1 : B$

4. Compute $\mathbf{V}_k$ from the SVD of $\mathbf{\Omega}((k-1)M_k + 1 : kM_k,\ 1 : N_T)$

5. Solve the triangular system $\mathbf{RF}_k = \mathbf{V}_k(:,\ 1 : M_k)$

6. End

The Per-user SLNR precoding based on GSVD computation produces better results than all the conventional methods. However, the objective function based on per-user SLNR neglects to take the intra-user antenna interference into account. Hence, this formulation is sub-optimum for spatial multiplexing gain extraction.

4. The proposed precoding by maximization of per-antenna signal-to-leakage-plus-noise ratio

The precoding technique originally proposed in [65] maximizes the SLNR for each user, thus the precoder so designed just cancels the inter-user interference. The technique proposed herein, however, utilizes a new cost function that seeks to maximize the Per-Antenna Signal-to-Leakage-plus-Noise Ratio (PA-SLNR) which would help minimize even the intra-user antenna interference. Thus, the precoder so designed maximizes the overall SLNR per user more efficiently. This is justified because the PA-SLNR as explained in figure 5, takes into account the intra-user antenna interference cancelation. For j^{th} receive antenna of kth user, the PA-SLNR given by γ_k^j, is defined as the ratio between the desired signal power of jth receive antenna to the interference introduced by the signal power intended for j^{th} antenna but leaked to all other antennas plus the noise power at that receiving antenna front end. So for the jth receive antenna of kth user, the PA-SLNR, γ_k^j is defined by:

$$\gamma_k^j = \frac{\| \mathbf{h}_k^j \mathbf{f}_k^j \|_F^2}{\sum_{\substack{i=1 \\ i \neq k}}^{B} \| \mathbf{H}_i \mathbf{f}_k^j \|_F^2 + \sum_{\substack{i=1 \\ i \neq j}}^{M_k} \| \mathbf{h}_k^i \mathbf{f}_k^j \|_F^2 + \sigma_{n_k^j}^2} \tag{29}$$

where $\mathbf{h}_k^j \in \mathbf{C}^{1\times N_T}$ is the kth user, j^{th} antenna received row. If we define an auxiliary matrix $\mathbf{H}_k^j$ as the matrix that contains all the received antennas rows of kth user except the jth row as follows:

$$\mathbf{H}_k^j = \begin{bmatrix} h_k^{(1,1)} & h_k^{(1,2)} & \cdots & h_k^{(1,N_T)} \\ \vdots & \vdots & \vdots & \vdots \\ h_k^{(j-1,1)} & h_k^{(j-1,2)} & \cdots & h_k^{(j-1,N_T)} \\ h_k^{(j+1,1)} & h_k^{(j+1,2)} & \cdots & h_k^{(j+1,N_T)} \\ \vdots & \vdots & \vdots & \vdots \\ h_k^{(M_k,1)} & h_k^{(M_k,2)} & \cdots & h_k^{(M_k,N_T)} \end{bmatrix} \in \mathbf{C}^{((M_k-1)\times N_T)} \tag{30}$$

and the combined channel matrices for all other systems receive antennas except the jth desired receive antenna row as as:

$$\widetilde{\mathbf{H}}_k^j = [\mathbf{H}_k^{jT} \mathbf{H}_1^T \cdots \mathbf{H}_{k-1}^T \mathbf{H}_{k+1}^T \cdots \mathbf{H}_B^T]^T \tag{31}$$

then from equation (31) and equation (30) the optimization expression in (29) can be rewritten as:

$$\gamma_k^j = \frac{\| \mathbf{h}_k^j \mathbf{f}_k^j \|_F^2}{\| \widetilde{\mathbf{H}}_k^j \mathbf{f}_k^j \|_F^2 + \sigma_{\mathbf{v}_k^j}^2} \tag{32}$$

Problem Formulation: For any j^{th} receive antenna of the kth user, select the precoding vector $\mathbf{f}_k^j$, where $k=1, \cdots, B$, $j=1, \cdots, M_k$ such that the PA-SLNR ratio is maximized:

$$\begin{aligned} &\mathbf{f}_k^j = \arg\max_{\mathbf{f}_k^j \in \mathbf{C}^{N_T\times 1}} \frac{\mathbf{f}_k^{j*}(\mathbf{h}_k^{j*}\mathbf{h}_k^j)\mathbf{f}_k^j}{\mathbf{f}_k^{j*}(\widetilde{\mathbf{H}}_k^{j*}\widetilde{\mathbf{H}}_k^j + \sigma_{n_k^j}^2 \mathbf{I}_{N_T})\mathbf{f}_k^j} \\ &\text{subject to:} \\ &\quad \mathrm{tr}(\mathbf{F}_k \mathbf{F}_k^*) = 1 \\ &\mathbf{F}_K = [\mathbf{f}^1, \cdots, \mathbf{f}^{M_k}] \end{aligned} \tag{33}$$

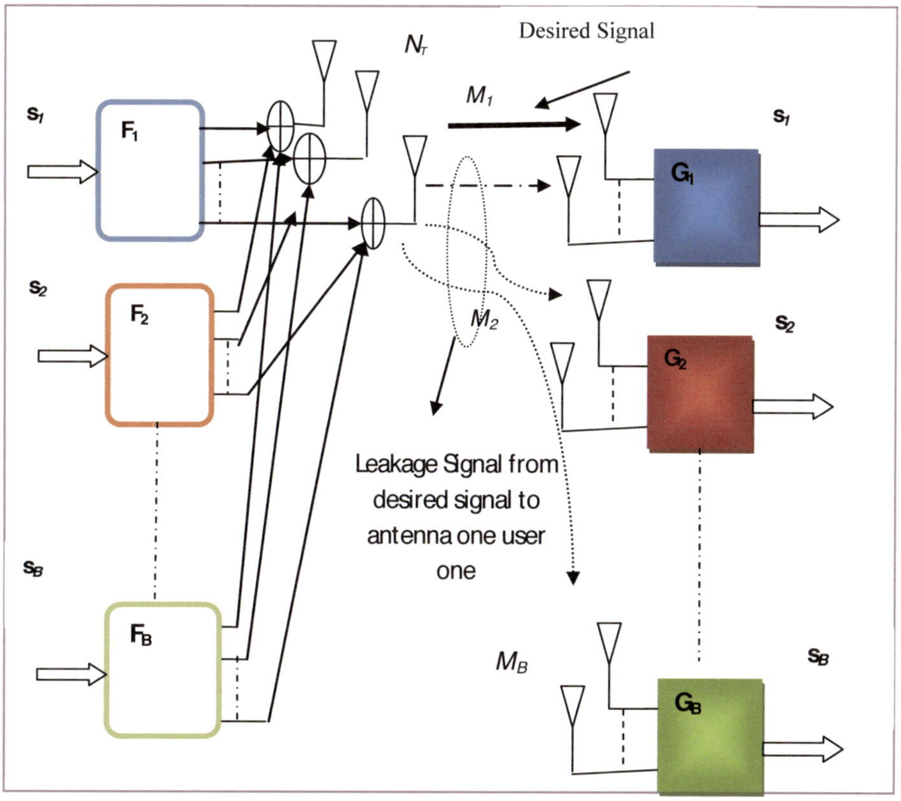

Figure 5. System Model depicts all Variables.

The optimization problem in the equation (33) deals with the jth antenna desired signal power in the numerator and a combination of total leaked power from desired signal to the jth antenna to all other antennas plus noise power at the jth antenna front end in the denominator. To calculate the precoding matrix for each user we need to calculate the precoding vector for each receive antenna independently. This requires solving the linear fractional optimization problem in the equation (33) $M_k \times B$ times using either GEVD [65] or GSVD [66] which both leading-to high computational load at the base stations. In the next section, Fukunaga-Koontz Transform (FKT) based solution method for solving such series of linear fractional optimization problems is described and simple computational method for MU-MIMO precoding algorithm is developed.

4.1. FKT and FKT based precoding algorithm

Fukunaga-Koontz Transform (FKT) is a normalization transform process which was first introduced in [67] to extract the important features for separating two pattern classes in pattern recognition. Since the time it was first introduced, FKT is used in many Linear Discriminant Analysis (LDA) applications notably in [68, 69]. Researchers in [70, 71] formulate the problem of recognition of two classes as follows: Given the data matrices ψ_1 and ψ_2,then from these two classes, the autocorrelation matrices $\mathbf{\Pi}_1 = \psi_1\psi_1^T$ and $\mathbf{\Pi}_2 = \psi_2\psi_2^T$ are positive semi-definite (p.s.d) and symmetric. For any given p.s.d autocorrelation matrices $\mathbf{\Pi}_1$ and $\mathbf{\Pi}_2$ the sum $\mathbf{\Pi}$ is still p.s.d and can be written as:

$$\mathbf{\Pi} = \mathbf{\Pi}_1 + \mathbf{\Pi}_2 = \begin{bmatrix} \mathbf{U} & \mathbf{U}_\perp \end{bmatrix} \begin{bmatrix} \mathbf{D} & \mathbf{0} \\ \mathbf{0} & \mathbf{0} \end{bmatrix} \begin{bmatrix} \mathbf{U}^T \\ \mathbf{U}_\perp^T \end{bmatrix} \tag{34}$$

Without loss of generality the sum $\mathbf{\Pi}$ can be singular and $r = rank(\mathbf{\Pi}) \le \mathbf{D}im(\mathbf{\Pi})$, where $\mathbf{D} = \text{diag}\,(\lambda_1, \cdots \lambda_r)$ and also $\lambda_1 \ge \cdots \ge \lambda_r > 0$.$\mathbf{U} \in \mathbf{C}^{\text{Dim}(\mathbf{\Pi}) \times r}$ is the set of eigenvectors that correspond to the set of nonzero-eigenvalues and $\mathbf{U}_\perp \in \mathbf{C}^{\text{Dim}(\mathbf{\Pi}) \times (\text{Dim}(\mathbf{\Pi}) - r)}$ is the orthogonal complement of $\mathbf{U}$. From the equation (34), the FKT transformation [71] matrix operator is defined as:

$$\mathbf{P} = \mathbf{U}\mathbf{D}^{-1/2} \tag{35}$$

By using this FKT transformation factor, the sum p.s.d matrix $\mathbf{\Pi}$ can be whitened such that the sum of the two Sub-matrices $\tilde{\mathbf{\Pi}}_1$ and $\tilde{\mathbf{\Pi}}_2$ gives the identity matrix as follows:

$$\mathbf{P}^T \mathbf{\Pi} \mathbf{P} = \mathbf{P}^T (\mathbf{\Pi}_1 + \mathbf{\Pi}_2) \mathbf{P} = \tilde{\mathbf{\Pi}}_1 + \tilde{\mathbf{\Pi}}_2 = \mathbf{I}^{r \times r} \tag{36}$$

Where $\tilde{\mathbf{\Pi}}_1 = \mathbf{P}^T \mathbf{\Pi}_1 \mathbf{P}$, $\tilde{\mathbf{\Pi}}_2 = \mathbf{P}^T \mathbf{\Pi}_2 \mathbf{P}$ are the transformed covariance matrices for $\mathbf{\Pi}_1$ and $\mathbf{\Pi}_2$ respectively, and $\mathbf{I}^{r \times r}$ is an identity matrix. Suppose that $\mathbf{v}$ is an eigenvector of $\tilde{\mathbf{\Pi}}_1$ with corresponding eigenvalue λ_1, then $\tilde{\mathbf{\Pi}}_1 \mathbf{v} = \lambda_1 \mathbf{v}$ and from the equation (36) we have $\tilde{\mathbf{\Pi}}_1 = \mathbf{I} - \tilde{\mathbf{\Pi}}_2$. Thus, the following results can be pointed:

$$(\mathbf{I} - \tilde{\mathbf{\Pi}}_2)\mathbf{v} = \lambda_1 \mathbf{v} \tag{37}$$

$$\tilde{\mathbf{\Pi}}_2 \mathbf{v} = (1 - \lambda_1)\mathbf{v} \tag{38}$$

This means that $\tilde{\mathbf{\Pi}}_2$ has the same eigenvectors as $\tilde{\mathbf{\Pi}}_1$ with corresponding eigenvalues related as $\lambda_2 = (1 - \lambda_1)$. Thus, we can conclude that the dominant eigenvectors of $\tilde{\mathbf{\Pi}}_1$ is the weakest eigenvectors of $\tilde{\mathbf{\Pi}}_2$ and vice versa. Based on the FKT transform analysis we conclude that the transformed matrices $\tilde{\mathbf{\Pi}}_1$ and $\tilde{\mathbf{\Pi}}_2$ share the same eigenvectors and the sum of the two corresponding eigenvalues are equal to one. Thus, the following decomposition is valid for any positive definite and positive semi-definiate matrices:

$$\tilde{\mathbf{\Pi}}_1 = \mathbf{V}\mathbf{\Lambda}_1\mathbf{V}^T \tag{39}$$

$$\tilde{\mathbf{\Pi}}_2 = \mathbf{V}\mathbf{\Lambda}_2\mathbf{V}^T \tag{40}$$

$$\mathbf{I} = \mathbf{\Lambda}_1 + \mathbf{\Lambda}_2 \tag{41}$$

Where $\mathbf{V} \in \mathbf{C}^{r\times r}$ is the matrix that contains all the eigenvectors, and $\mathbf{\Lambda}_1$, $\mathbf{\Lambda}_2$ are the corresponding eigenvalues matrices. Thus, from these analyses we conclude that FKT gives the best optimum solution for any fractional linear problem without going through any serious matrix inversion step. By relating FKT transform analysis of the two covariance matrices, and the precoding design problem for MU-MIMO (multiple linear fractional optimization problem), we can make direct mapping of the optimization variable from equation (33) to the FKT transform as follows:

$$\mathbf{\Pi}_1 = \mathbf{h}_k^{j*}\mathbf{h}_k^j \tag{42}$$

$$\mathbf{\Pi}_2 = \tilde{\mathbf{H}}_k^{j*}\tilde{\mathbf{H}}_k^j + \sigma^2{}_k^j\mathbf{I}_{N_T} \tag{43}$$

and consequently the sum $\mathbf{\Pi}$ of the two covariance matrices becomes:

$$\mathbf{\Pi} = \mathbf{H}_{\mathrm{com}}^{*}\mathbf{H}_{\mathrm{com}} + \sigma^2{}_k^j\mathbf{I}_{N_T} \tag{44}$$

where $\mathbf{H}_{com}$ is the combined channel matrix for all user which is given as $\mathbf{H}_{com} = [\mathbf{H}_1^T\,\mathbf{H}_2^T \cdots \mathbf{H}_B^T]^T$.

According to FKT analysis, we can calculate the FKT factor and consequently we use this transformation factor to generate the shared eigenspace matrices $\tilde{\mathbf{\Pi}}_1$ and $\tilde{\mathbf{\Pi}}_2$ using the facts from equations (42-44) for each receive antenna in the system. The shared eigen subspaces are complements of each other such that the best principal eigenvectors of the first transformed covariance matrix $\tilde{\mathbf{\Pi}}_1$ are the least principal eigenvector for the second transformed covariance matrix $\tilde{\mathbf{\Pi}}_2$ and vice-versa. Thus, we can find the receive antenna precoding vector by simply multiplying the FKT factor with the eigenvectors corresponding to the best eigenvalue of the transformed antenna covariance matrix or eigenvector corresponding to the least eigenvalue of the transformed leakage plus noise covariance matrix. The most notable observation is that, for the set of $M_K \times B$ receiving antennas in the system, we need to compute the FKT transform factor only once, which cuts down the computation load sharply. Algorithm 2, summarizes the computation steps of the precoding matrix for multiple B users in the system using FKT.

Algorithm 2: PA-SLNR MU-MIMO precoding based on FKT for multiple B independent MU-MIMO users.

• Input: Combined channel matrix for all Busers and the input noise variance

$\mathbf{H}_{com} = [\mathbf{H}_1^T \mathbf{H}_2^T \cdots \mathbf{H}_B^T]^T, \sigma_k^2$

• Output: Precoding matrices $\mathbf{F}_k$ for multiple Busers such that, $k = 1, \cdots, B$

1. Compute the sum $\mathbf{\Pi} = \mathbf{H}_{com}^{*} \mathbf{H}_{com} + \sigma_k^{2j} \mathbf{I}_{N_T}$

2. Compute FKT factor $\mathbf{P} = \mathbf{U}\mathbf{D}^{-\frac{1}{2}}$ from $SVD(\mathbf{\Pi})$

3. For k= 1 to B

4. For j=1 to M_k

○ Transform the jth receive antenna covariance matrix $\mathbf{\Pi}_1$ using the FKT factor $\mathbf{P}$to $\tilde{\mathbf{\Pi}}_1$ and select the first eigenvector $\mathbf{v}_k^j$ of $\tilde{\mathbf{\Pi}}_1$

○ The precoding vector corresponding to the j^{th}receive antenna at the kth user is: $\mathbf{f}_k^j = \mathbf{P}\mathbf{v}_k^j$

End

○ Synthesize the kth user precoding matrix is $\mathbf{F}_k = [\mathbf{f}_k^1 \cdots \mathbf{f}_k^{M_k}]$

End

The algorithm takes the combined MU-MIMO channel matrix as well as the value of the noise variance as an input and outputs B users precoding matrices. It computes the FKT factor in step one and two and iterates B times (step three to six) to calculate the precoding matrices for B number of users. For each user, there are M_k sub-iteration operations (step four to five) to calculate each individual user precoding matrix in vector by vector basis.

4.2. Performance evaluation

In this section we will highlight the importance of Precoding for MU-MIMO and showcase the performance of the proposed PA-SLNR-FKT scheme in two scenarios – scenario 1, single cell MU-MIMO where there is no interference from other cell and scenario 2 of multi-cell processing (MCP) where there is multicell interference and the objective is to make use of multiple antennas in all basestation cooperatively to improve the overall system performance. We use both basic assumptions and a typical simplified WiMAX physical layer Standard discrete channel Models for the Monte-Carlo simulation. Firstly, with basic assumptions we provide comparative performance evaluation results of the proposed MU-MIMO Precoder and the PU-SLNR maximization techniques using GEVD and GSVD proposed in [34, 58]. The comparison is done in terms of average received BER and output received SINR outage performance metrics. In each simulation setup, the entries of kth user MIMO channel $\mathbf{H}_k$ is generated as complex white Gaussian random variables with zero mean and unit variance. The users data

Parameter	Configuration
System configuration	MU-MIMO-BC
Each user channel	Matrix elements generated as zero-mean and unit-variance i.i.d complex Gaussian random variables
Modulation	4-QAM
Precoding methods	• Proposed method (PA-SLNR-FKT) • Reference SLNR-GEVD, SLNR-GSVD
Performance metrics	Received BER and Received SINR outage
MIMO Decoder	Matched filter.
Number of Iterations	• 5000 System transmission for BER calculation • 2000 System transmission running for SINR outage calculation

Table 1. Narrowband MU-MIMO System Configuration Summary

symbol vectors are modulated and spatially multiplexed at the base station. At the receivers, matched filter is used to decode each user's data. Detailed summary of the MU-MIMO-BC system configuration parameters are given in table (1).

4.2.1. Scenario 1: Single cell MU-MIMO

In this scenario we consider single cell transmission where implicitly we assume that the multi-cell interference is zero. Figure 6. shows the average received BER performance of the proposed PA-SLNR-FKT and the reference methods of SLNR-GSVD and SLNR-GEVD precoding schemes for the MU-MIMO-BC system configurations of $N_T = 14$, $B = 3M_k = 4$. In this configuration, the numbers of the base station antennas are more than the sum of all receiving antennas which also signifies more degree of freedom in MU-MIMO transmission. In this simulation also, the base station utilizes 4-QAM modulation to modulate, spatially multiplexes and precodes a vector of length 4 symbols to each user. The average BER is calculated over 5000 MU-MIMO channel realization for each algorithm. The proposed method outperforms SLNR-GSVD and SLNR-GEVD. At BER equal to 10^{-4} there is approximately 4dB performance gain over SLNR-GSVD.

Figure 7 compares the received output SINR outage performance of the proposed PA-SLNR-FKT precoding and the reference SLNR-GSVD and SLNR-GEVD precoding methods. MU-MIMO system with full rate configuration of $B = 3$, $M_k = 4$, $N_T = 12$ and 2dB input SNR is considered in the simulation. The proposed method outperforms the SLNR-GSVD method by 1 dB gain at 10% received output SINR outage.

Figure 8 compares the received output SINR outage performance of the proposed PA-SLNR-FKT precoding and the reference SLNR-GSVD and SLNR-GEVD precoding methods. MU-MIMO system with full rate configuration of $B = 3$, $M_k = 4$, $N_T = 12$ and 10dB input SNR are considered in the simulation. The proposed method outperforms the SLNR-GSVD method by approximately 1.5 dB gain at 10% received output SINR outage.

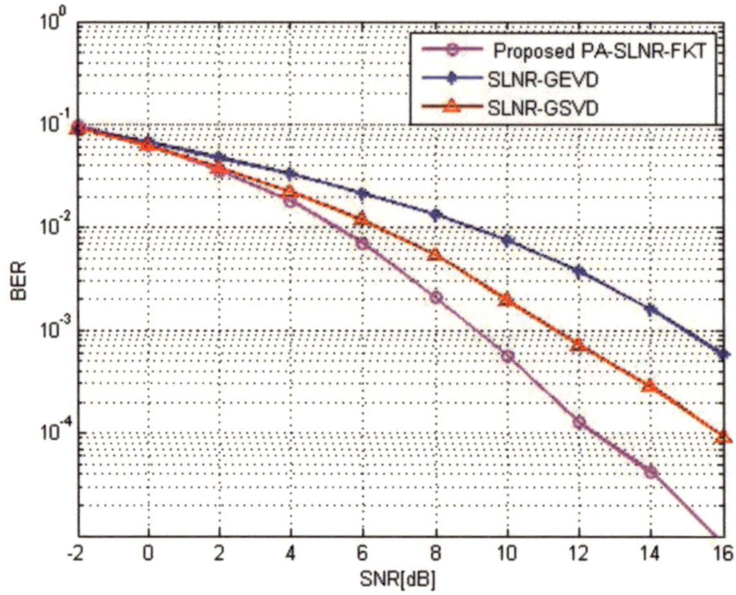

Figure 6. Shows The Compression of The Un-coded BER Performance for PA-SLNR-FKT, SLNR-GSVD and SLNR-GEVD Precoding Methods Under System Configuration of $B=3$, $M_k=4$, $N_T=14$, 4QAM Modulated Signal.

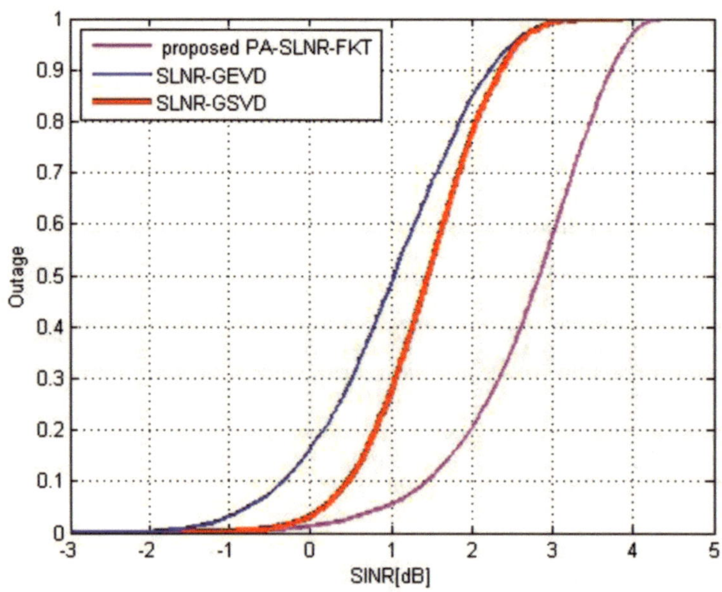

Figure 7. Shows The Comparison of The Output SINR Outage Performance of The PA-SLNR-FKT, SLNR-GSVD and SLNR-GEVD Precoding Methods Under System Configuration of $B=3$, $M_k=4$, $N_T=12$, and 2 dB Input SNR.

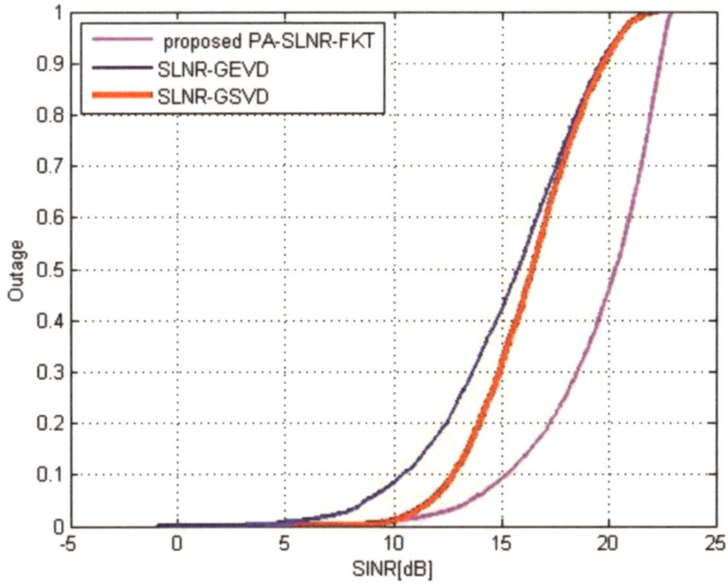

Figure 8. Shows The Comparison of the Output SINR Outage Performance PA-SLNR-FKT, SLNR-GSVD and SLNR-GEVD Precoding Methods Under System Configuration of $B=3$, $M_k=4$, $N_T=12$, and 10 dB Input SNR.

4.2.2. Scenario 2: MU-MIMO with Multi Cell Processing (MCP)

To appreciate the importance of precoding to cancel inter-cell interference in the MCP configuration, geometrically we consider three cooperating BSs as illustrated in figure 9. Basically we consider micro-cellular setup of BS to BS distance equal to 1000 m. As shown in figure 9. we also consider a simplified as well as an extreme case where there are three users at the edges of the three cooperating cells, so that we just allow each user to uniformly position within the last 50 m of its anchor BS. In each transmission, a WiMAX standard channel model is used. Thus each entry of the kth user MIMO channel matrix is generated according to pre-specified wireless communication channel model which include mean path loss, shadowing and slow fading discrete components as follows.

$$\mathbf{H}_e^k = (\phi_1\phi_2)^{1/2}\ddot{\mathbf{H}}_e^k \tag{45}$$

where $\overleftrightarrow{\mathbf{H}}_e^k \in \mathbf{C}^{N_{R_k}\times N_{T_e}}$ represent the fast fading channel discrete component between the kth user and the eth BS and in this system simulation we use the WiMAX discrete channel values as given in [72] and ϕ_1 denotes the channel path loss component while ϕ_2 is the lognormal shadowing fading component. In each step of simulation fixed least square filter is used to decode the received data and unless specified otherwise, the following values listed in table 2. are used.

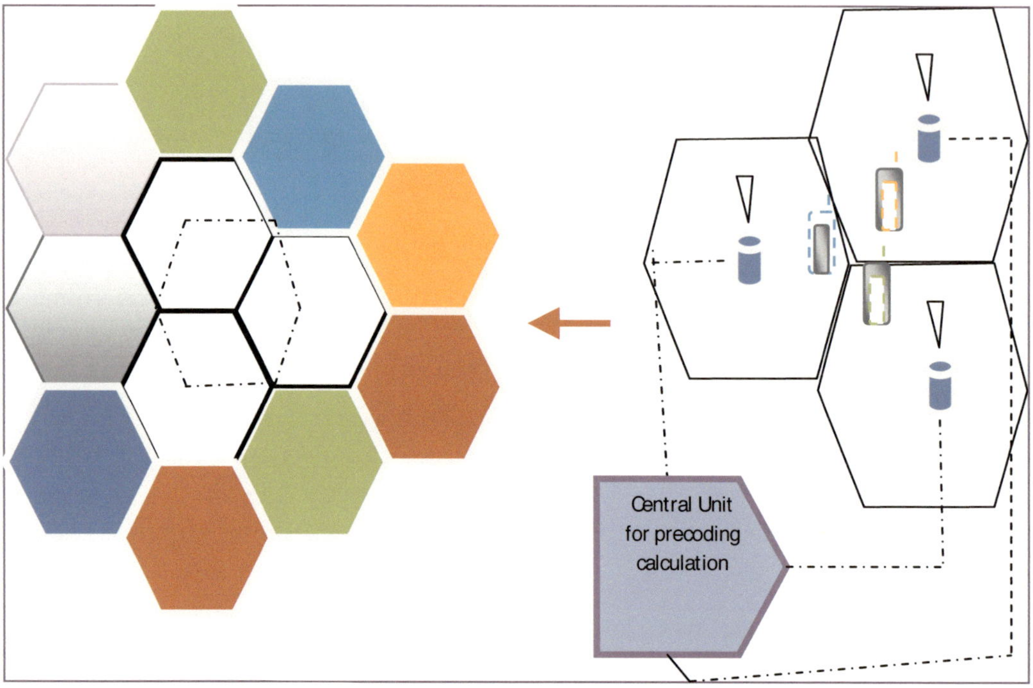

Figure 9. Example of Tidy Three Multi-cells Cooperation (Exchanging Both the Data and the CSI)

Channel Parameter	Parameter Value
$\phi_1 = \varepsilon d^{\alpha}$ where ε, d α are the intercept, BS to MS distance and path loss exponent [72, 73].	$\varepsilon = 1.35 \times 10^7$, $\alpha = -3$
$\phi_2 = 10^{\xi/10}$	ξ is generated as zero mean real value Gaussian random variable with standard deviation equal to 8dB
Signal to noise ratio at the cell edge	(2-20) dB
The number of simulation runs	1000 simulations running
System configuration notations	E denotes the number of cooperated cells. N_T denotes the number of transmit antennas at each cells. $N_{\mathbf{R}}$ denotes the number of receive antennas at each end user. B denotes the number of user.

Table 2. System Simulation Parameters.

Figure 10. shows the simulated average user rate performance for the MU-MIMO MCP. In this figure, the proposed algorithm result is denoted by PA-SLNR-FK-MCP and the conventional precoding methods of PU-SLNR-GEVD-MCP and PU-SLNR-GSVD-MCP and as a benchmark we also consider the case of no cooperation denoted as NO-MCP. The simulated configuration is for $E=3.B=3$, $N_T=5$, $N_{\mathbf{R}}=4$, with 1000 WiMAX discrete channel realizations. It is also clearly shown that without MCP there is no valuable rate for the cell edge user. Partially these results also support the claim that the proposed method has better performance than the related works at 18dB input SNR, there is approximately 0.5 bits/s/Hz sum rate gain over SLNR-GSVD.

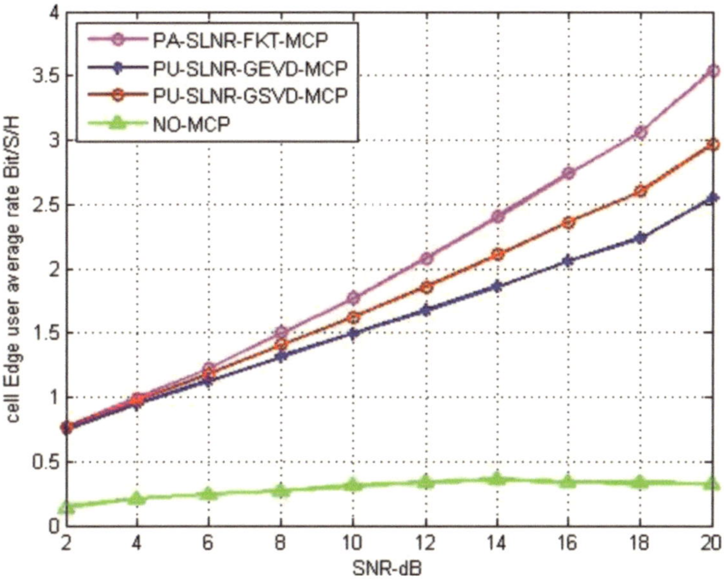

Figure 10. Average Cell Edge User Rate Performance of MCP Precoding Methods for the Configuration ($E=3$, $N_T=5$, $N_{\mathbf{R}}=4$, $B=3$), WiMAX Channel Model Precoding for Multi-cell Processing (Networked MIMO)

As shown before, the simulation result reveals the unique opportunities arising from MU-MIMO transmission optimization of antenna spatial multiplexing/ spatial diversity techniques with Multi-cell Processing (Inter-cell interference cancelation). Furthermore, it also clearly indicates that MU-MIMO precoding approaches provide significant multiplexing (on the order of the number of antennas used at the transmitter) and diversity gains while resolving some of the issues associated with conventional cellular systems. Particularly, it brings precoding robustness with MU-MIMO gain and turn the inter-cell interference into diversity.

5. Conclusion and future research directions

To conclude, this chapter has introduced the principles of MIMO techniques, reviewed various MU-MIMO precoding methods, and extended the knowledge by proposing a new method that outperformed the methods available in the literature. The results as shown in in figure10, demonstrated conclusively the significant role of precoding in inter-cell interference cancelation in MCP scenario. There are still some interesting open issues and topics for future research related to this application. For example related to the data and CSI sharing, basically we assume perfect system interconnection in TDD mode of operation but how much system pilot is needed for the multi-cell cooperation is an open problem. Also we assume that there is no system error or delay. This assumption is an ideal assumption for typical system deployment. The real conditions are non-ideal, therefore, modeling and investigation of the effects of system errors and delay are also an open issue that need to be researched.

Author details

Elsadig Saeid, Varun Jeoti and Brahim B. Samir

Electrical and Electronic Engineering Department, Universiti Teknologi PETRONAS, Tronoh, Perak, Malaysia

References

[1] Telatar, "Capacity of Multi-antenna Gaussian Channel," *European Transactions on Telecommunications,* vol. 10, pp. 585-595, October 1995.

[2] G. J. Foschini, "Layered space-time architecture for wireless communication in a fading environment when using multi-element antennas," *Bell Labs Technical Iournals,* vol. 1, pp. 41-59, 14 AUG 1996.

[3] J. Winters, "On the capacity of radio communication systems with diversity in rayleigh fading environment," *IEEE Journal on Selected Areas in Communications,* vol. 5, pp. 871-878, June 1987.

[4] S. M. Alamouti, "A simple transmit diversity scheme for wireless communications," *IEEE Journal on Selected Areas in Communications,* vol. 16, pp. 1451-1458, 06 August 1998.

[5] P. DeBeasi. (2008, 20 Aug). *802.11n: Enterprise Deployment Considerations*. Available: http://www.wi-fi.org/

[6] F. Khalid and J. Speidel, "Advances in MIMO techniques for mobile communications- Asurvey," *Int'l J. of Communications, Network and System Sciences,* vol. 3, pp. 213-252, March 2010.

[7] A. Goldsmith, *Wirless Communications,* First ed. Cambridge: Cambridge University Press., 2005.

[8] D. Tse and P. Viswanath, *Fundamentals of Wireless Communication,* Frist ed. Cambridge: Cambridge university Press, 2005.

[9] S. Da-Shan, *et al.*, "Fading correlation and its effect on the capacity of multielement antenna systems," *IEEE Transactions on Communications,* vol. 48, pp. 502-513, Mar 2000.

[10] A. B. Gershman and N. D. Sidiropoulos, *Space-time processing for MIMO communications*, First ed.: Johon wiley & Sons, Ltd, 2005.

[11] D.-s. Shiu, *et al.*, "fading correlation and Its effect on the capacity of multielement antenna systems," *IEEE TRANSACTION On ComunicationS,* vol. Vol. 48, pp. 502-512, March 2000.

[12] A. Paulraj, *et al.*, *Introduction to Space-Time Wireless Communications*, First ed. New York: Cambridge university press, 2003.

[13] R. Gallager. course materials for 6.450 Principles of Digital Communications I, [Online]. Available: http//ocw.mit.edu

[14] T. M. Cover and J. A. Thomas, *Elements of Information Theory*, Second ed.: JOHN WILEY & SONS, INC.,, 2006.

[15] H. Sampath, *et al.*, "Generalized Linear Precoder and Decoder Design for MIMO Channels Using The weighted MMSE Criterion," *IEEE TRANSACTION On Comunications,* vol. 49, pp. 2198-2206, DEC 2001.

[16] A. Lozano and N. Jindal, "Transmit Diversity vs. Spatial Multiplexing in Modern MIMO Systems," *IEEE TRANSACTIONS ON WIRELESS COMMUNICATIONS,* vol. 9, pp. 186-197, 08 January 2010.

[17] E. A. Lee and D. G. Messerschmitt, *Digital Communication*, second ed. Boston: Kluwer, 1994.

[18] Y. S. Cho, *et al.*, *MIMO-OFDM Wireless Communications With Matlab*, First ed.: John Wiley & Sons, 2010.

[19] J. W. Huang, *et al.*, "Precoder Design for Space-Time Coded MIMO Systems with Imperfect Channel State Information," *IEEE Transactions on Wireless Communications,* vol. 7, pp. 1977-1981, 2008.

[20] L. Zhi-Quan, *et al.*, "Transceiver optimization for block-based multiple access through ISI channels," *IEEE Transactions on Signal Processing,* vol. 52, pp. 1037-1052, 2004.

[21] P. Mary, *et al.*, "Symbol Error Outage Analysis of MIMO OSTBC Systems over Rice Fading Channels in Shadowing Environments," *IEEE transactions on wireless communications,* vol. 10, pp. 1009 - 1014 15 April 2011.

[22] A. Goldsmith, *et al.*, "Capacity limits of MIMO channels," *IEEE Journal on Selected Areas in Communications,* vol. 21, pp. 684-702, 2003.

[23] A. Saad, *et al.*, "Capacity of MIMO channels at different antenna configurations," *Journal of Applied Sciences,* vol. 8, pp. 4595-4602, 2008.

[24] M. Trivellato, *et al.*, "On channel quantization and feedback strategies for multiuser MIMO-OFDM downlink systems," *IEEE Transactions ON Comm,* vol. 57, pp. 2645-2654, Sept 2009.

[25] M. Costa, "Writing on dirt paper," *IEEE TRANSACTIONS ON Information theory,* vol. 29, pp. 439-441, May 1983.

[26] H. Sampath, *et al.*, "Generalized linear precoder and decoder design for MIMO channels using the weighted MMSE criterion," *IEEE Transactions on Communications,* vol. 49, pp. 2198-2206, 2001.

[27] D. J. Love, *et al.*, "Grassmannian beamforming for multiple-input multiple-output wireless systems," *IEEE Transactions on Information Theory,* vol. 49, pp. 2735-2747, 2003.

[28] Z. Jianchi, *et al.*, "Investigation on precoding techniques in E-UTRA and proposed adaptive precoding scheme for MIMO systems," in *14th Asia-Pacific Conference on Communications, 2008. APCC 2008.* , 2008, pp. 1-5.

[29] R. Narasimhan, "Spatial multiplexing with transmit antenna and constellation selection for correlated MIMO fading channels," *IEEE Transactions on Signal Processing,* vol. 51, pp. 2829-2838, 2003.

[30] V. Stankovic, "Multi-user MIMO wireless communication," PhD, Technischen Universit°at Ilmenau, Ilmenau, 2006.

[31] M. Costa, "Writing on dirty paper (Corresp.)," *IEEE Transactions on Information Theory,* vol. 29, pp. 439-441, 1983.

[32] E. Saeid, *et al.*, "Efficient Per-antenna Signal to Leakage Plus noise Ratio Precoding for Multiuser Multiple Input Multiple Output System," *Research Journal of Applied Sciences, Engineering and Technology,* vol. 4, pp. 2489-2495,, August 2012.

[33] V. Stankovic and M. Haardt, "Generalized Design of Multi-User MIMO Precoding Matrices," *Wireless Communications, IEEE Transactions on,* vol. 7, pp. 953-961, 2008.

[34] M. Sadek, *et al.*, "A Leakage-Based Precoding Scheme for Downlink Multi-User MIMO Channels," *IEEE TRANSACTION On Comunications,* vol. 6, pp. 1711-1721, May 2007.

[35] C. B. Peel, *et al.*, "A vector-perturbation technique for near-capacity multiantenna multiuser communication-part I: channel inversion and regularization," *IEEE Transactions on Communications,* vol. 53, pp. 195-202, 2005.

[36] V. Mai and A. Paulraj, "MIMO Wireless Linear Precoding," *Signal Processing Magazine, IEEE,* vol. 24, pp. 86-105, 2007.

[37] A. Kurve, "Multi-user MIMO systems: the future in the making," *IEEE Potentials,* vol. 28, pp. 37-42, 2009.

[38] C. Wang, *et al.*, "On the Performance of the MIMO Zero-Forcing Receiver in the Presence of Channel Estimation Error," presented at the Information Sciences and Interaction Sciences, Chengdu, China 2007.

[39] Y. Jiang, *et al.*, "Performance Analysis of ZF and MMSE Equalizers for MIMO Systems: An In-Depth Study of the High SNR Regime," *IEEE TRANSACTIONS ON Information theory,* vol. 57, pp. 2008 - 2026 april 2011.

[40] A. Sibille, *et al., MIMO From Theory to Implementation,* First ed.: Imprint: Academic Press, 2011.

[41] L. Zheng, "Diversity-Multiplexing Tradeo: A Comprehensive View of Multiple Antenna Systems," PhD, Electrical Engineering and Computer Sciences, CALIFORNIA at BERKELEY, 2002.

[42] Z. Lizhong and D. N. C. Tse, "Diversity and multiplexing: a fundamental tradeoff in multiple-antenna channels," *Information Theory, IEEE Transactions on,* vol. 49, pp. 1073-1096, 2003.

[43] S. Ahmadi, "An overview of next-generation mobile WiMAX technology," *Communications Magazine, IEEE,* vol. 47, pp. 84-98, 2009.

[44] K. Etemad, "Overview of mobile WiMAX technology and evolution," *Communications Magazine, IEEE,* vol. 46, pp. 31-40, 2008.

[45] C. Swannach and G. W. Wonrnel, "Channel State Quantization IN MIMO Broadcast System: Architectures and Codes," PhD, Electrical Elgineering and Computer Science, MIT, 2010.

[46] S. Fang, *et al.*, "Multi-User MIMO Linear Precoding with Grassmannian Codebook," in *WRI International Conference on Communications and Mobile Computing, 2009. CMC '09.* , 2009, pp. 250-255.

[47] M. C. H. Lim, *et al.*, "Spatial Multiplexing in the Multi-User MIMO Downlink Based on Signal-to-Leakage Ratios," in *Global Telecommunications Conference, 2007. GLOBECOM '07. IEEE,* 2007, pp. 3634-3638.

[48] S. Shi, "Transceiver Design for Multiuser MIMO Systems," Phd, Electrical and Electronic Engineering, Berlin, Berlin, 2009.

[49] S. S. Christensen, *et al.*, "Weighted sum-rate maximization using weighted MMSE For MIMO-BC Beamforming Design," *IEEE TRANS . On . wireless Comm.,* vol. vol. 7 pp. 1-7, Dec 2008.

[50] X. Gao, *et al.*, "A successive iterative optimization precoding method for downlink multi-user MIMO system," in *International Conference on Wireless Communications and Signal Processing (WCSP),* 2010, pp. 1-5.

[51] A. J. Tenenbaum and R. S. Adve, "Linear Processing and sum throughput in the Multiuser MIMO downlink," *IEEE Trans On wireless Comm.*, vol. 8, pp. 2652-2660, May 2009.

[52] V. Stankovic, "Iterative Successive MMSE Multi-User MIMO Transmit Filtering," *ELEC. ENERG*, vol. 20, pp. 45-55, April 2007.

[53] M. Joham, *et al.*, "Linear transmit processing in MIMO communications systems," *IEEE Transactions on Signal Processing*, vol. 53, pp. 2700-2712, 2005.

[54] M. Joham, *et al.*, "Transmit Wiener filter for the downlink of TDDDS-CDMA systems," in *IEEE Seventh International Symposium on Spread Spectrum Techniques and Applications* 2002, pp. 9-13

[55] L. Min and O. Seong Keun, "A Per-User Successive MMSE Precoding Technique in Multiuser MIMO Systems," in *Vehicular Technology Conference, 2007. VTC2007-Spring. IEEE 65th*, 2007, pp. 2374-2378.

[56] H. Karaa, *et al.*, "Linear Precoding for Multiuser MIMO-OFDM Systems," in *IEEE International Conference on Communications*, 2007, pp. 2797-2802.

[57] V. Sharma and S. Lambotharan, "Interference Suppression in Multiuser Downlink MIMO Beamforming using an Iterative Optimization Approach," presented at the 14th European Signal Processing Conference, Florence, Italy, 2006.

[58] J. Park, *et al.*, "Generalised singular value decomposition-based algorithm for multiuser multiple-input multiple-output linear precoding and antenna selection," *IET Communcation*, vol. 4, pp. 1899–1907 5 November 2010.

[59] M. Sadek, "Transmission Techniques for Multi-user MIMO communications," Phd, Electrical Engineering, University of California, LA, 2006.

[60] P. Jaehyun, *et al.*, "Efficient GSVD Based Multi-User MIMO Linear Precoding and Antenna Selection Scheme," in *IEEE International Conference on Communications* 2009, pp. 1-6.

[61] C. Peng, *et al.*, "A New SLNR-Based Linear Precoding for Downlink Multi-User Multi-Stream MIMO Systems," *Communications Letters, IEEE*, vol. 14, pp. 1008-1010, 2010.

[62] C. Paige and M. A. Saunders, "Towards a Generalized Singular Value Decomposition," *SIAM Journal on Numerical Analysis*, vol. 18, pp. 398-405, 1981.

[63] C. F. and V. Loan, "Generalized Singular value Decomposition," *SIAM Journal on Numerical Analysis* vol. 13, March 1976.

[64] G. H. Golub and C. F. v. V. Loan, *Matrix Computations*, 3rd Edition ed. Baltimore and London The Johns Hopkins University Press 1996.

[65] M. Sadek, *et al.*, "Active antenna selection in multiuser MIMO communications," *IEEE TRANS ON SIGNAL PROCESSING*, vol. 44, pp. 1498-1510, April 2007.

[66] J. Park, *et al.*, "Efficient GSVD Based Multi-user MIMO Linear Precoding and Antenna Selection Scheme," in *IEEE ICC 2009*, Dresden 2009, pp. 1-6.

[67] K. Fukunaga and W. L. G. Koontz, "Application of the Harhunen-loeve expansion to feature selection and ordering," *IEEETransaction onn Computer,* vol. 19, pp. 311-317, April 1970.

[68] A. A. Miranday and P. F. Whelan, "Fukunaga-Koontz transform for small sample size problems," in *ISSC 2005 - IEE Irish Signals and Systems Conference*, Dublin., 2005, pp. 1-6.

[69] W. Cao and R. Haralick, "Affine feature extraction: A Generalization of the Fukunaga-Koontz transformation," in *international conference on Machine Learning and Data Mining in Pattern Recognition* Heidelberg, 2007, pp. 163-173.

[70] Z. Sheng and T. Sim, "Discriminant Subspace Analysis: A Fukunaga-Koontz Approach," *Pattern Analysis and Machine Intelligence, IEEE Transactions on,* vol. 29, pp. 1732-1745, 2007.

[71] S. Zhang and T. Sim, "When Fisher meets Fukunaga-Koontz: A New Look at Linear Discriminants," in *IEEE Computer Society Conference on Computer Vision and Pattern Recognition (CVPR'06)*, New York, 2006, pp. 323 - 329

[72] G. L. Stüber, *Principles of Mobile Communication*, Second ed.: Kluwer Academic Publishers, 2002.

[73] V. Erceg, *et al.*, "An empirically based path loss model for wireless channels in suburban environments," *Selected Areas in Communications, IEEE Journal on,* vol. 17, pp. 1205-1211, 1999.

Chapter 2

PAPR Reduction in WiMAX System

Mona Shokair and Hifzalla Sakran

Additional information is available at the end of the chapter

http://dx.doi.org/10.5772/55380

1. Introduction

Broadband Wireless Access (BWA) has emerged as a promising solution for last mile access technology to provide high speed internet access in the residential as well as small and medium sized enterprise sectors. At this moment, cable and DSL technologies are providing broadband service in these sectors. But the practical difficulties in deployment have prevented them from reaching many potential broadband internet customers. Many areas throughout the world currently are not under broadband access facilities. Even many urban and suburban locations may not be served by DSL connectivity as it can only reach about three miles from the central office switch [1]. On the other side many older cable networks do not have return channel which will prevent to offer internet access and many commercial areas are often not covered by cable network. But with BWA this difficulties can be overcomed. Because of its wireless nature, it can be faster to deploy, easier to scale and more flexible, thereby giving it the potential to serve customers not served or not satisfied by their wired broadband alternatives.

IEEE 802.16 standard for BWA and its associated industry consortium, Worldwide Interoperability for Microwave Access (WiMAX) forum promise to offer high data rate over large areas to a large number of users where broadband is unavailable. This is the first industry wide standard that can be used for fixed wireless access with substantially higher bandwidth than most cellular networks [2]. Wireless broadband systems have been in use for many years, but the development of this standard enables economy of scale that can bring down the cost of equipment, ensure interoperability, and reduce investment risk for operators.

The first version of the IEEE 802.16 standard operates in the 10–66GHz frequency band and requires line of sight (LOS) towers. Later the standard extended its operation through different PHY specification to 2-11 GHz frequency band enabling non line of sight (NLOS) connections, which require techniques that efficiently mitigate the impairment of fading and multipath [3]. Taking the advantage of OFDM technique the PHY is able to provide robust broadband service in hostile wireless channel. The OFDM based physical layer of the IEEE 802.16 standard has been standardized in close cooperation with the European Telecommunications Standards Institute (ETSI) High PERformance Metropolitan Area Network (HiperMAN) [4]. Thus, the HiperMAN standard and the OFDM based physical layer of IEEE 802.16 are nearly identical. Both OFDM based physical layers shall comply with each other and a global OFDM system should emerge [5]. The WiMAX forum certified products for BWA comply with the both standards.

Some researchers investigate the effect of nonlinear amplifier in WiMAX [6] and apply Clipping as a simple method [7], CORDIC Algorithm [8], and Tone Reservation [9] to reduce PAPR in WiMAX system.

In this chapter, proposed WiMAX system will be studied. This system will be compared with the conventional system. Where the companding technique is used to reduce PAPR based on the properties of the μ-law that uses for decreasing dynamics range of the signal. Moreover, the performance of the proposed system will be compared with the system that uses clipping as a reduction of PAPR. These systems will be investigated under SUI channels and AWGN.

In Section 1.2, the broadband channel will be explained. In section 1.3 SUI multipath Channel Models will be investigated. The performance of MMSE equalizer over SUI model will be study in section 1.4. In section 1.5, a system model of WiMAX for PAPR will be explained. PAPR Reduction Technique Using μ-Law Compander is studied in Section 1.6. Finally, summary will be made.

2. Broadband wireless channel models

One of the more intriguing aspects of wireless channels is fading. Unlike path loss or shadowing, which are large-scale attenuation effects owing to distance or obstacles, fading is caused by the reception of multiple versions of the same signal. The multiple received versions are caused by reflections that are referred to as *multipath*. The reflections may arrive nearly simultaneously— for example, if there is local scattering around the receiver—or at relatively longer intervals— for example, owing to multiple paths between the transmitter and the receiver (Figure 1).

When some of the reflections arrive at nearly the same time, their combined effect is as in Figure 2. Depending on the phase difference between the arriving signals, the interference can be either constructive or destructive, which causes a very large observed difference in the amplitude of the received signal even over very short distances. In other words, moving the transmitter or the receiver even a very short distance can have a dramatic effect on the received amplitude, even though the path loss and shadowing effects may not have changed at all.

One of the key parameters in the design of a transmission system is the maximum delay spread value that it has to tolerate.

In order to design and benchmark wireless communication systems, it is important to develop channel models that incorporate their variations in time, frequency, and space. Models are

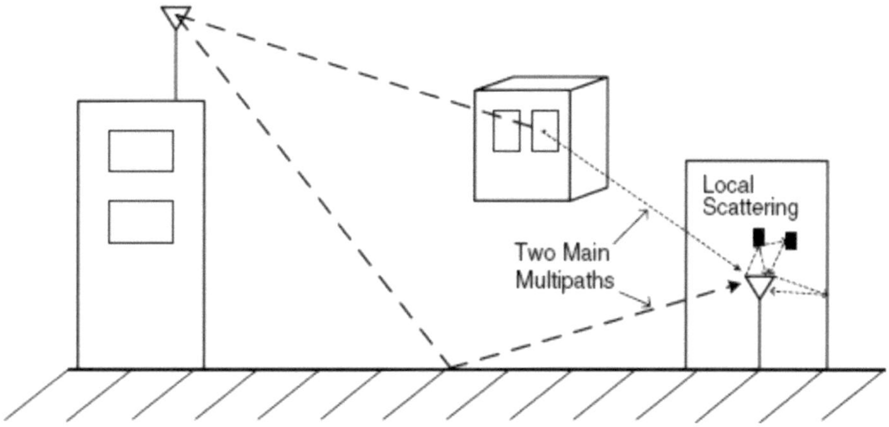

Figure 1. A channel with a few major paths of different lengths, with the receiver seeing a number of locally scattered versions of those paths.

classified as either *statistical* or *empirical*. Statistical models are simpler and are useful for analysis and simulations. Empirical models are more complicated but usually represent a specific type of channel more accurately. There are several channels models which are explained in the following,

2.1. Statistical channel models

As we have noted, the received signal in a wireless system is the superposition of numerous reflections, or multipath components. The reflections may arrive very closely spaced in time — for example, if there is local scattering around the receiver—or at relatively longer intervals.

Figure 2 shows that when the reflections arrive at nearly the same time, constructive and destructive interference between the reflections causes the envelope of the aggregate received signal *r*(t) to vary substantially.

In this section, we summarize statistical methods for characterizing the amplitude and power of *r*(t) when all the reflections arrive approximately at the same time.

2.1.1. Rayleigh fading

Suppose that the number of scatters is large and that the angles of arrival between them are uncorrelated. From the Central Limit Theorem, it can be shown that the in-phase (cosine) and quadrature (sine) components of *r*(t), denoted as $r_I(t)$ and $r_Q(t)$, follow two independent time correlated Gaussian random processes.

Consider a snapshot value of at time t =0, and note that $r(0)= r_I(0) + r_Q(0)$. Since the values $r_I(0)$ and $r_Q(0)$ are Gaussian random variables, it can be shown that the distribution of the envelope

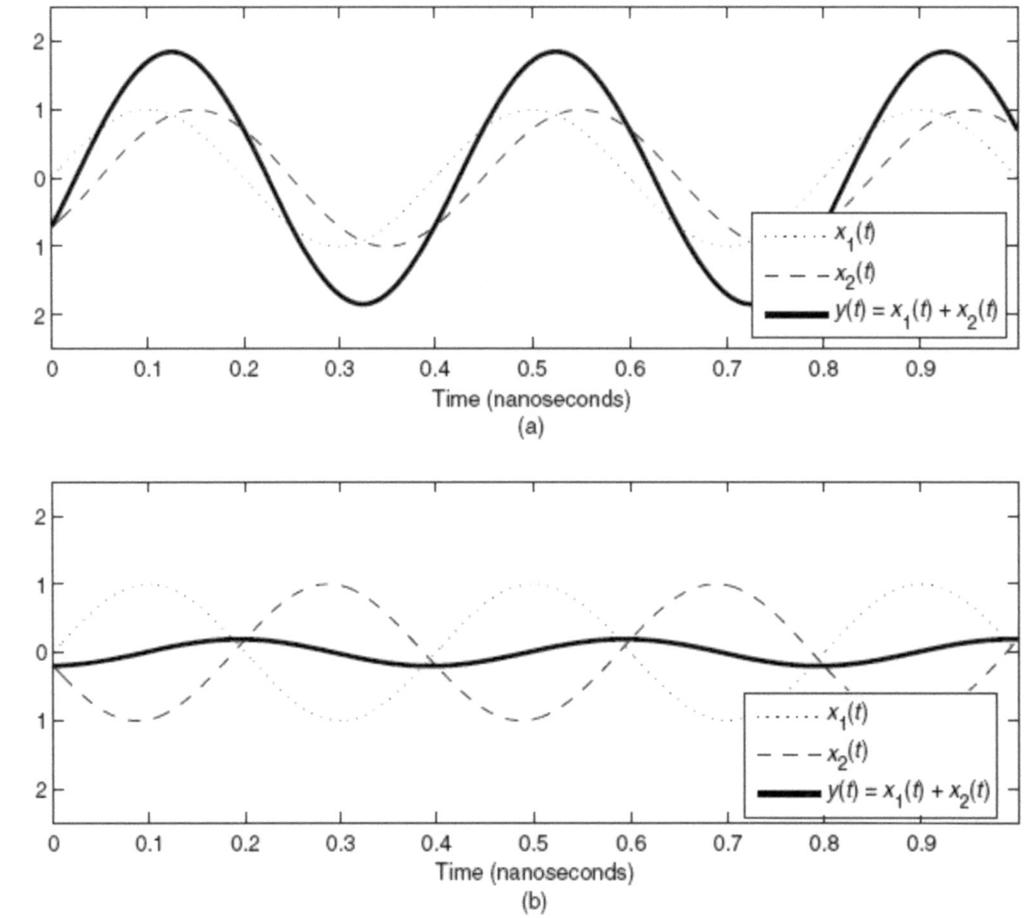

Figure 2. The difference between (a) constructive interference and (b) destructive interference at *fc* = 2.5GHz is less than 0.1 nanoseconds in phase, which corresponds to about 3 cm.

amplitude $|r| = \sqrt{r_I^2 + r_Q^2}$ is Rayleigh and that the received power $|r|^2 = r_I^2 + r_Q^2$ is exponentially distributed. Formally [10],

$$f_{|r|}(x) = \frac{2x}{p_r} e^{-x^2/p_r} \quad , \quad x \geq 0, \tag{1}$$

and

$$f_{|r|^2}(x) = \frac{2x}{p_r} e^{-x/p_r} \quad , \quad x \geq 0, \tag{2}$$

where p_r is the average received power.

2.1.2. LOS channels: Ricean distribution

An important assumption in the Rayleigh fading model is that all the arriving reflections have a mean of zero. This will not be the case if there is a dominant path—for example, a LOS path — between the transmitter and the receiver. For a LOS signal, the received envelope distribution is more accurately modeled by a Ricean distribution, which is given by

$$f_{|r|}(x) = \frac{x}{\sigma^2} e^{-(x^2+\mu^2)/2\sigma^2} I_0\left(\frac{x\mu}{\sigma^2}\right) \quad , \quad x \geq 0, \tag{3}$$

where μ^2 is the power of the LOS component, σ^2 is the variance and I_0 is the 0th-order, modified Bessel function of the first kind. Although more complicated than a Rayleigh distribution, this expression is a generalization of the Rayleigh distribution. This can be confirmed by observing that

$$\mu = 0 \Rightarrow I_0\left(\frac{x\mu}{\sigma^2}\right) = 1 \quad ,$$

Therefore the Ricean distribution reduces to the Rayleigh distribution in the absence of a LOS component.

Since the Ricean distribution depends on the LOS component's power μ^2, a common way to characterize the channel is by the relative strengths of the LOS and scattered paths. This factor, K, is quantified as

$$K = \frac{\mu^2}{2\sigma^2}$$

and is a natural description of how strong the LOS component is relative to the NLOS components.
For $K=0$, the Ricean distribution again reduces to Rayleigh, and as $K \to \infty$, the physical meaning is that there is only a single LOS path and no other scattering. Mathematically, as K grows large, the Ricean distribution is quite Gaussian about its mean μ with decreasing variance, physically meaning that the received power becomes increasingly deterministic.

The average received power under Ricean fading is the combination of the scattering power and the LOS power: $P_r = 2\sigma^2 + \mu^2$. Although it is not straightforward to directly find the Ricean power distribution $f_{|r|^2}(x)$, the Ricean envelope distribution in terms of K can be found by substituting $\mu^2 = KP_r / (K+1)$ and $2\sigma^2 = P_r / (K+1)$ into Equation (3).

Although its simplicity makes the Rayleigh distribution more amenable to analysis than the Ricean distribution, the Ricean distribution is usually a more accurate depiction of wireless broadband systems, which typically have one or more dominant components. This is especially true of fixed wireless systems, which do not experience fast fading and often are deployed to maximize LOS propagation.

2.2. Empirical channel models

The parametric statistical channel models discussed therefore in this chapter do not take into account specific wireless propagation environments. Although exactly modeling a wireless channel requires complete knowledge of the surrounding scatterers, such as buildings and plants, the time and computational demands of such a methodology are unrealistic, owing to the near-infinite number of possible transmit/receive locations and the fact that objects are subject to movement. Therefore, empirical and semiempirical wireless channel models have been developed to accurately estimate the path loss, shadowing, and small-scale fast fading. Although these models are generally not analytically tractable, they are very useful for simulations and to fairly compare competing designs. Empirical models are based on extensive measurement of various propagation environments, and they specify the parameters and methods for modeling the typical propagation scenarios in various wireless systems.

2.3. Stanford University Interim (SUI) channel models

SUI channel models are an extension of the earlier work by AT&T Wireless and Erceg. In this model a set of six channels was selected to address three different terrain types that are typical of the continental US [11]. This model can be used for simulations, design, development and testing of technologies suitable for fixed broadband wireless applications [12]. The parameters for the model were selected based upon some statistical models. The tables below depict the parametric view of the six SUI channels.

TerrainType	SUIChannels
C (Mostly flat terrain with light tree densities)	SUI1, SUI2
B (Hilly terrain with light tree density or flat terrain with moderate to heavy tree density)	SUI3, SUI4
A (Hilly terrain with moderate to heavy tree density)	SUI5, SUI6

Table 1. Terrain type for SUI channel.

The parametric view of the SUI channels is summarized in the Table 2. For simplicity, SUI 1 from train type C(Flat/Light tree density), SUI 3 from train type B(Hilly/Light tree density or Flat/moderate tree density),and SUI 5 from train type A(Hilly/moderate to heavy tree density) are considered in the following.

Model	Delay	L(numberoftape)=3		RMS Delay spread
Gain / Kfactor	Tap1	Tap2	Tap3	
SUI 1	0 μs	0.4 μs	0.9 μs	0.111 μs
	0 dB	-15 dB	-20 dB	
	4	0	0	
SUI 2	0 μs	0.4 μs	1.1 μs	0.202 μs
	0 dB	-12 dB	-15 dB	
	2	0	0	
SUI 3	0 μs	0.4 μs	0.9 μs	0.264 μs
	0 dB	-5 dB	-10 dB	
	1	0	0	
SUI 4	0 μs	1.5 μs	4 μs	1.257 μs
	0 dB	-4 dB	-8 dB	
	0	0	0	
SUI 5	0 μs	4 μs	10 μs	2.842 μs
	0 dB	-5 dB	-10 dB	
	0	0	0	
SUI 6	0 μs	14 μs	20 μs	5.240 μs
	0 dB	-10 dB	-14 dB	
	0	0	0	

Table 2. SUI Channels Parameter.

3. Equalization in wireless channel

Equalization defines any signal processing technique used at the receiver to alleviate the ISI problem caused by delay spread. Signal processing can also be used at the transmitter to make the signal less susceptible to delay spread: spread spectrum and multicarrier modulation fall in this category of transmitter signal processing techniques. ISI mitigation is required when the modulation symbol time T_s is on the order of the channel's rms delay spread σ_{Tm}. Higher data rate applications are more sensitive to delay spread, and generally require high-performance equalizers or other ISI mitigation techniques. In fact, mitigating the applications are more sensitive to delay spread, and generally require high-performance equalizers or other ISI mitigation techniques. In fact, mitigating the applications are more sensitive to delay spread, and generally require high-performance equalizers or other ISI mitigation techniques. In fact, mitigating the impact of delay spread is one of the most challenging hurdles for high-speed wireless data systems.

Equalizer design must typically balance ISI mitigation with noise enhancement, since both the signal and the noise pass through the equalizer, which can increase the noise power. Nonlinear equalizers suffer less from noise enhancement than linear equalizers, but typically entail higher complexity.

Equalization techniques fall into two broad categories: linear and nonlinear. The linear techniques are generally the simplest to implement and to understand conceptually. However, linear equalization techniques typically suffer from more noise enhancement than nonlinear equalizers.

Among nonlinear equalization techniques, decision-feedback equalization (DFE) is the most common, since it is fairly simple to implement and generally performs well. However, on channels with low SNR, the DFE suffers from error propagation when bits are decoded in error, leading to poor performance. The optimal equalization technique is maximum likelihood sequence estimation (MLSE). Unfortunately, the complexity of this technique grows exponentially with the length of the delay spread, and is therefore impractical on most channels of interest.

However, the performance of the MLSE is often used as an upper bound on performance for other equalization techniques.

It is clear that equalization in OFDM can be very simple. This is one of the major advantages of using OFDM over single carrier systems. Channel equalization in OFDM actually can be done by just a simple division in the frequency domain. This is because the channel as a filter is convolved with the input signal in the time domain on transmission. This operation is equivalent to multiplication in the frequency domain and thus undoing the effects of the channel is just a division.

This section studies the performance of OFDM system over multipath SUI channels which are not clarified until now. Moreover, the performance of this system will be compared with the performance of the system with the frequency domain equalizer (FDE) using MMSE. Also this system will be investigated over AWGN.

3.1. System model

The OFDM implementation of multicarrier modulation is shown in Figure 3. The input data stream is modulated by a QAM modulator, resulting in a complex symbol stream $X[0], X[1], \ldots, X[N-1]$ of length N. This symbol stream is passed through a serial-to-parallel converter, whose output is a set of N parallel QAM symbols $X[0], \ldots, X[N-1]$ corresponding to the symbols transmitted over each of the subcarriers. Thus, the N symbols output from the serial-to-parallel converter are the discrete frequency components of the OFDM modulator output $s(t)$. In order to generate $s(t)$, these frequency components are converted into time samples by performing an inverse DFT on these N symbols, which is efficiently implemented using the IFFT algorithm. The IFFT yields the OFDM symbol consisting of the sequence $x[n] = X[0], \ldots, x[N-1]$ of length N, where

$$x(n) = \sum_{i=0}^{N-1} X(i) e^{j2\pi ni/N}, 0 \leq n \leq N-1. \tag{4}$$

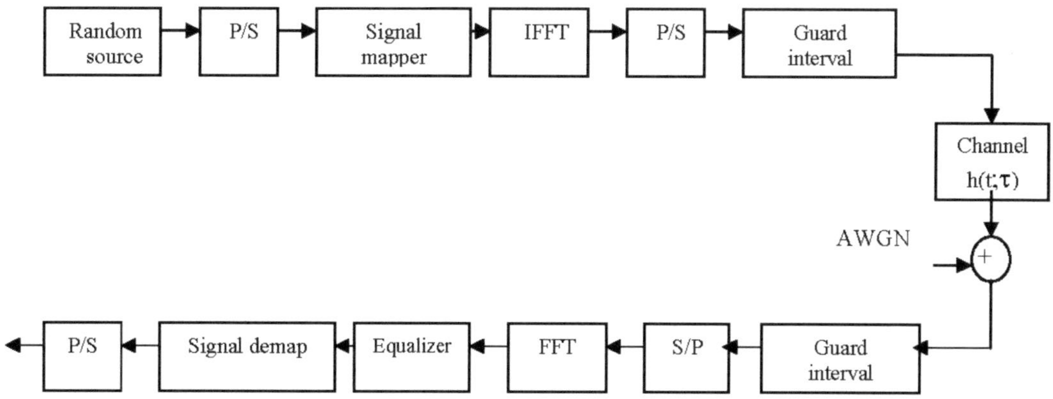

Figure 3. Model for studied OFDM-system.

This sequence corresponds to samples of the multicarrier signal: i.e. the multicarrier signal consists of linearly modulated subchannels, and the right hand side of (4) corresponds to samples of a sum of QAM symbols $X[i]$ each modulated by carrier frequency. The cyclic prefix is then added to the OFDM symbol, and the resulting time samples $\tilde{x}\,[n] = \tilde{x}[-\mu],\ldots,\tilde{x}\,[N-1] = x[N-\mu],\ldots, x[0],\ldots, x[N-1]$ are ordered by the parallel-to-serial converter[13].

The received signal is $r[n] = \tilde{x}[n] * h[n] + v[n], -\mu \le n \le N-1$.

Where $h(n)$ is impulse response of channel with length $\mu + 1 = T_m/T_s$, where T_m is the channel delay spread and T_s the sampling time associated with the discrete time sequence, $v[\mathrm{n}]$ is AWGN.

To simplify our derivation, we will choose N =8 subcarriers, prefix-length=2. Assume channel impulse response is : $h_0, h_1, h_2, 0, 0\ldots.$

Received samples for symbol m, after removing prefix:

$$\underbrace{\begin{bmatrix} r_m^0 \\ r_m^1 \\ r_m^2 \\ r_m^3 \\ r_m^4 \\ r_m^5 \\ r_m^6 \\ r_m^7 \end{bmatrix}}_{\text{received samples for symbol m}} = \underbrace{\begin{bmatrix} h_2 & h_1 & h_0 & 0 & 0 & 0 & 0 & 0 & 0 & 0 \\ 0 & h_2 & h_1 & h_0 & 0 & 0 & 0 & 0 & 0 & 0 \\ 0 & 0 & h_2 & h_1 & h_0 & 0 & 0 & 0 & 0 & 0 \\ 0 & 0 & 0 & h_2 & h_1 & h_0 & 0 & 0 & 0 & 0 \\ 0 & 0 & 0 & 0 & h_2 & h_1 & h_0 & 0 & 0 & 0 \\ 0 & 0 & 0 & 0 & 0 & h_2 & h_1 & h_0 & 0 & 0 \\ 0 & 0 & 0 & 0 & 0 & 0 & h_2 & h_1 & h_0 & 0 \\ 0 & 0 & 0 & 0 & 0 & 0 & 0 & h_2 & h_1 & h_0 \end{bmatrix}}_{\text{channel matrix}} \underbrace{\begin{bmatrix} x_m^6 \\ x_m^7 \\ x_m^0 \\ x_m^1 \\ x_m^2 \\ x_m^3 \\ x_m^4 \\ x_m^5 \\ x_m^6 \\ x_m^7 \end{bmatrix}}_{\text{transmitted sequence for symbol m}} \quad (5)$$

This is equivalent with:

$$\underbrace{\begin{bmatrix} r_m^0 \\ r_m^1 \\ r_m^2 \\ r_m^3 \\ r_m^4 \\ r_m^5 \\ r_m^6 \\ r_m^7 \end{bmatrix}}_{\text{received samples for symbol m}} = \underbrace{\begin{bmatrix} h_0 & 0 & 0 & 0 & 0 & 0 & h_2 & h_1 \\ h_1 & h_0 & 0 & 0 & 0 & 0 & 0 & h_2 \\ h_2 & h_1 & h_0 & 0 & 0 & 0 & 0 & 0 \\ 0 & h_2 & h_1 & h_0 & 0 & 0 & 0 & 0 \\ 0 & 0 & h_2 & h_1 & h_0 & 0 & 0 & 0 \\ 0 & 0 & 0 & h_2 & h_1 & h_0 & 0 & 0 \\ 0 & 0 & 0 & 0 & h_2 & h_1 & h_0 & 0 \\ 0 & 0 & 0 & 0 & 0 & h_2 & h_1 & h_0 \end{bmatrix}}_{\text{modified channel matrix}} . \begin{bmatrix} x_m^0 \\ x_m^1 \\ x_m^2 \\ x_m^3 \\ x_m^4 \\ x_m^5 \\ x_m^6 \\ x_m^7 \end{bmatrix} \tag{6}$$

The modified channel matrix is a so-called "circulant" matrix (constant along the diagonals & wrapped around). For every circulant matrix C is diagonalized by a DFT & I-DFT matrix:

$C = (IDFT).(\text{diagonal matrix}).(DFT)$

Diagonal matrix has DFT of first column of C on its main diagonal

By substituting this:

$$\underbrace{DFT. \begin{bmatrix} r_m^0 \\ r_m^1 \\ r_m^2 \\ r_m^3 \\ r_m^4 \\ r_m^5 \\ r_m^6 \\ r_m^7 \end{bmatrix}}_{\text{received symbol in freq.domain}} = \underbrace{\begin{bmatrix} H_0 & 0 & 0 & 0 & 0 & 0 & 0 & 0 \\ 0 & H_1 & 0 & 0 & 0 & 0 & 0 & 0 \\ 0 & 0 & H_2 & 0 & 0 & 0 & 0 & 0 \\ 0 & 0 & 0 & H_3 & 0 & 0 & 0 & 0 \\ 0 & 0 & 0 & 0 & H_4 & 0 & 0 & 0 \\ 0 & 0 & 0 & 0 & 0 & H_5 & 0 & 0 \\ 0 & 0 & 0 & 0 & 0 & 0 & H_6 & 0 \\ 0 & 0 & 0 & 0 & 0 & 0 & 0 & H_7 \end{bmatrix}}_{\text{channel matrix in freq.domain}} . \begin{bmatrix} X_m^0 \\ X_m^1 \\ X_m^2 \\ X_m^3 \\ X_m^4 \\ X_m^5 \\ X_m^6 \\ X_m^7 \end{bmatrix} \tag{7}$$

which means that after removing the prefix-samples and performing a DFT in the receiver, the obtained samples are equal to the transmitted (`frequency-domain') symbols, up to a channel attenuation H_i (for tone-i). Hence channel equalization may be performed in the frequency domain, by component-wise divisions (divide by H_i for tone-i) (1-taps FDE).

The multi-path fading channel can be written as:

$$r = H_c x + v \tag{8}$$

The channel equalization issue will be investigated in OFDM. Let *h*(t) designate the channel impulse response and H(w) its Fourier transform, i.e., the channel transfer function. If the number of carriers is sufficiently large, the channel transfer function becomes virtually nonselective within the bandwidth of each individual carrier. Focusing on one particular carrier, the influence of multipath fading reduces to attenuation and a phase rotation.

Referring back to the channel transfer function H(w), we let

$$\mathrm{H}_k = \rho_k . e^{j\theta_k} \tag{9}$$

Designate its value within the bandwidth of the *kth* carrier. Equalization of the channel requires that at the DFT output in the receiver, the *kth* carrier signal be multiplied by a complex coefficient

$$C_k = 1 / H_k \tag{10}$$

This is the result of an optimization based on the zero-forcing (ZF) criterion [14], which aims at canceling ISI regardless of the noise level. To minimize the combined effect of ISI and additive noise, the equalizer coefficients can be optimized under the minimum mean-square error (MMSE) criterion. This optimization yield

$$\mathrm{C}_k = \frac{\mathrm{H}_k^*}{\left|\mathrm{H}_k\right|^2 + \sigma_n^2 / \sigma_a^2} \tag{11}$$

Where σ_n^2: is the variance of additive noise, and σ_a^2 is the variance of the transmitted data symbols. Note that the MMSE solution reduces to the ZF solution for of $\sigma_n^2 = 0$.

Channel equalization in OFDM systems thus takes the form of a complex multiplier bank at the DFT output in the receiver.

The ZF criterion does not have a solution if the channel transfer function has spectral nulls in the signal bandwidth. Inversion of the channel transfer function requires an infinite gain and leads to infinite noise enhancement at those frequencies corresponding to spectral nulls. In general, the MMSE solution is more efficient, as it makes a trade-off between residual ISI (in the form of gain and phase mismatchs) and noise enhancement. This is particularly attractive for channels with spectral nulls or deep amplitude depressions.

3.2. Frequency-domain equalization

Analyzing the operation principle of OFDM, Frequency domain equalization is illustrated in Figure 4a which shows the baseband equivalent model of a single-carrier system employing this equalization technique. The received signal samples are passed to an N-point DFT, each output sample is multiplied by a complex coefficient C_i, and the output is passed to an IDFT to transform the signal back to the time domain. Now, if we take the system sketched in Figure 4a and place it between an IDFT operator and a DFT operator, we obtain an OFDM system incorporating a frequency domain equalizer. Obviously, the DFT and IDFT operators at the output end cancel each other, and the system simplifies to what we see in Figure 4b. This is precisely the schematic diagram of the equalized OFDM system discussed in the previous. Figures 4a and 4b give evidence of the strong similarities of OFDM signaling and frequency domain equalization in single-carrier systems. In cases, time/ frequency and frequency/ time transformations are made. The difference is that in OFDM systems, both channel equalization and receiver decisions are performed in the frequency domain, whereas in single-carrier systems the receiver decisions are made in the time domain although channel equalization is performed in the frequency domain.

In this section we study the performance MMSE equalizer rather than ZF over SUI MultiPath Channels, because the MMSE solution is more efficient [15]-[17], as it makes a trade-off between residual ISI (in the form of gain and phase mismatchs) and noise enhancement.

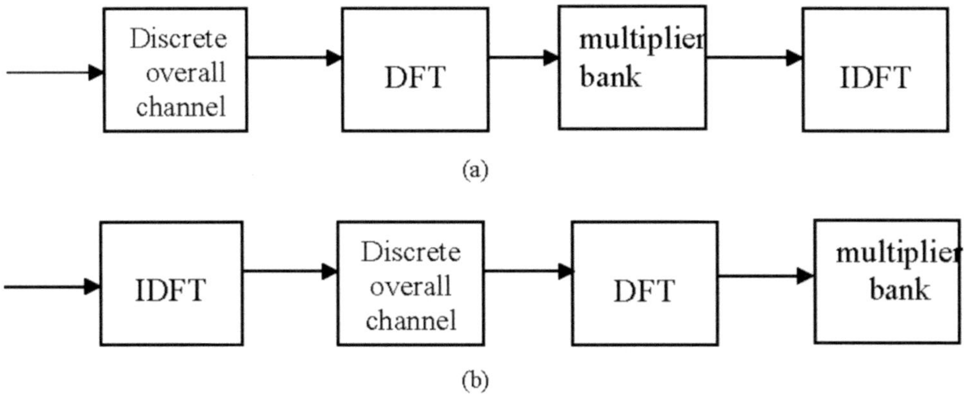

Figure 4. Frequency domain channel equalization. (a) Single carrier (b) OFDM

3.3. Simulation results

To compare the performance of OFDM systems in frequency selective fading channel, we consider OFDM block transmission over SUI channel model which is multipahs channel adopted by IEEE 802.16a task group for evaluating Broadband wireless system in 2-11 GHz bands. The sampling rate was assumed 20 Ms/sec.

This simulated system employs an OFDM signal with N=256, and 512 sub carriers using QPSK, 16 QAM, and 64 QAM, respectively.

In this simulation, SUI 1 from train type C(Flat/Light tree density), SUI3 from train type B(Hilly/ Light tree density or Flat/moderate tree density), and SUI 5 from train type A(Hilly/moderate to heavy tree density) are assumed for simplicity.

BER performance of OFDM system using MMES equalizer will be investigated in Figures 5, 6, 7, 8, 9,and 10 over SUI 1,3,5 channel models and AWGN for N=256 and 512 at 16QAM, 64 QAM and QPSK, respectively. Moreover, the performance of OFDM over SUI channel will be compared with the AWGN only as shown in Figures 11 and 12 for N=256 and N=512, respectively.

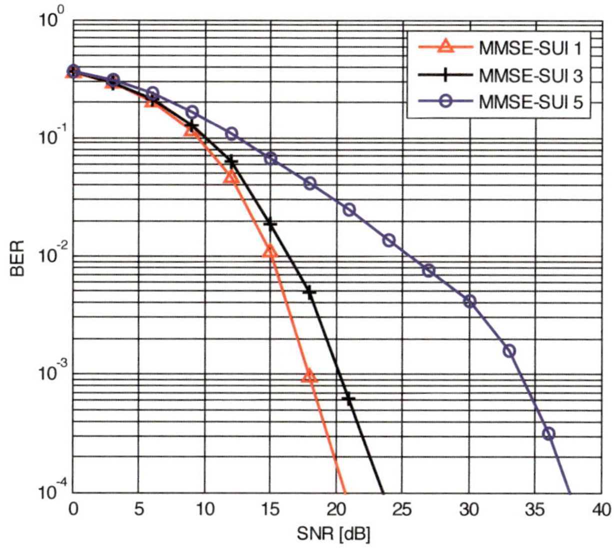

Figure 5. OFDM BER Performance (N=256, 16QAM)

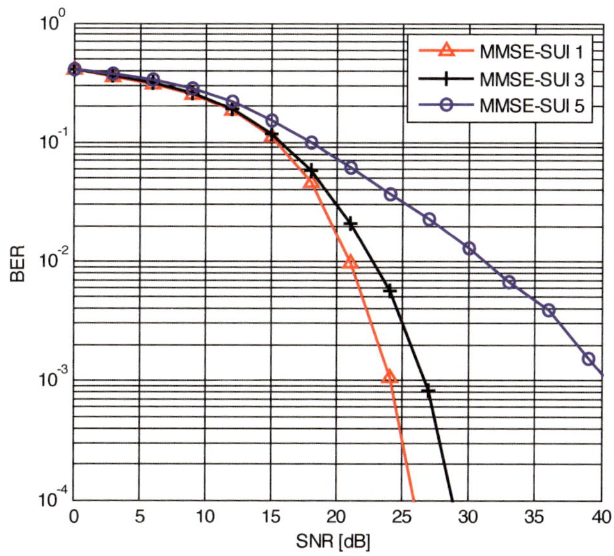

Figure 6. OFDM BER Performance (N=256, 64QAM)

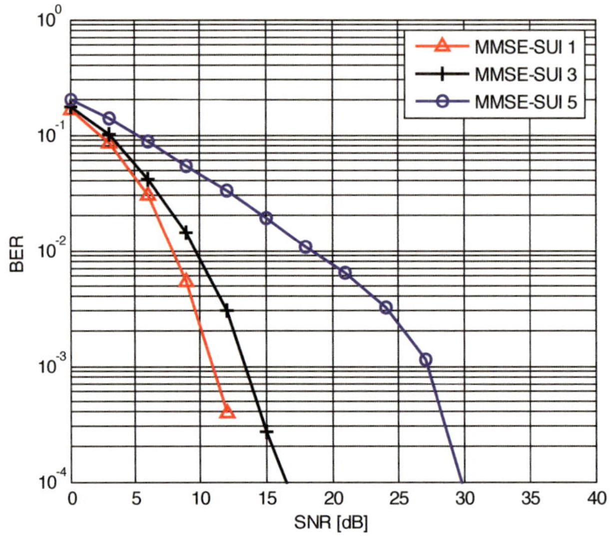

Figure 7. OFDM BER Performance (*N*=256, QPSK)

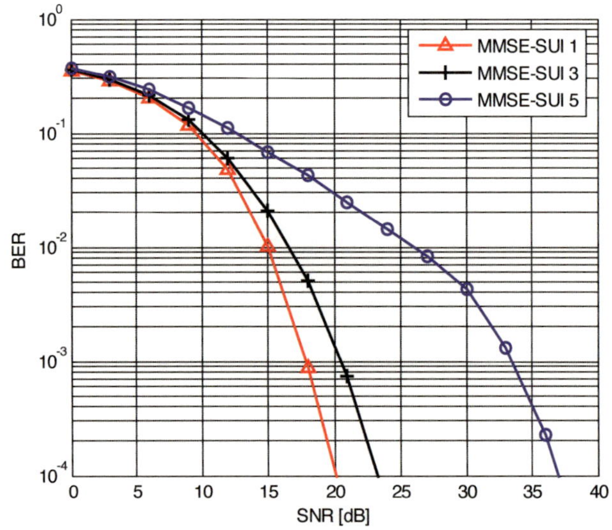

Figure 8. OFDM BER Performance (*N*=512, 16QAM)

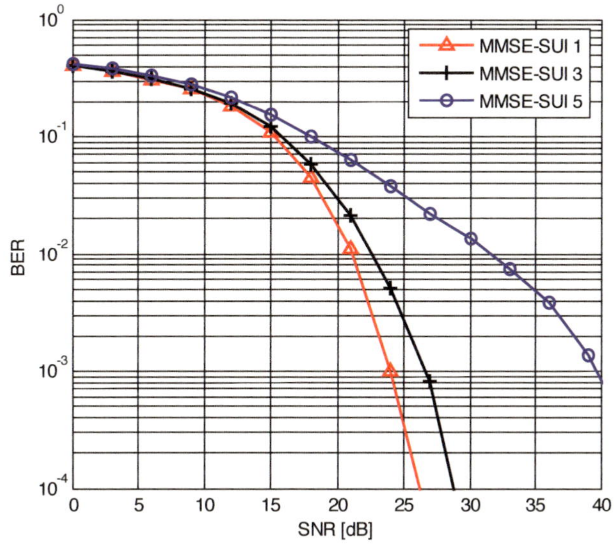

Figure 9. OFDM BER Performance (*N*=512, 64QAM)

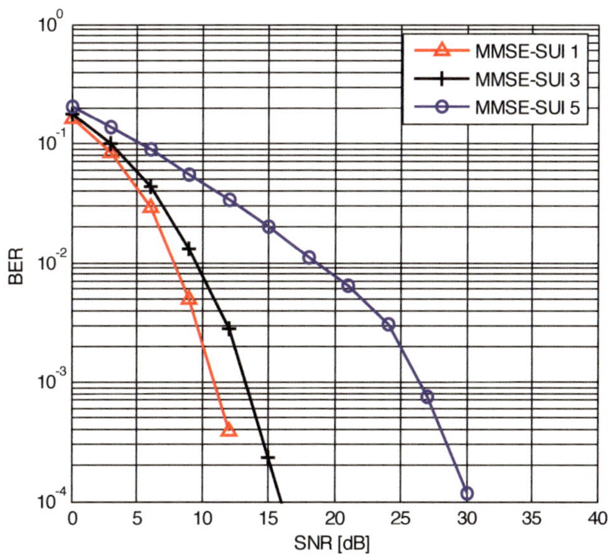

Figure 10. OFDM BER Performance (*N*=512, QPSK)

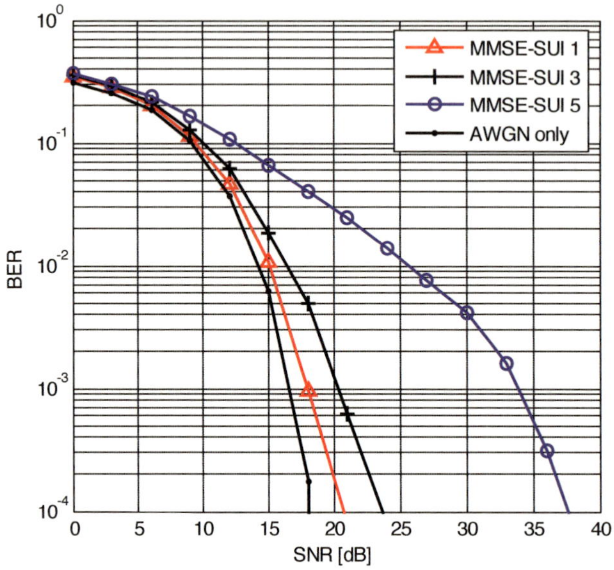

Figure 11. BER Performance comparison (16QAM, *N*=256).

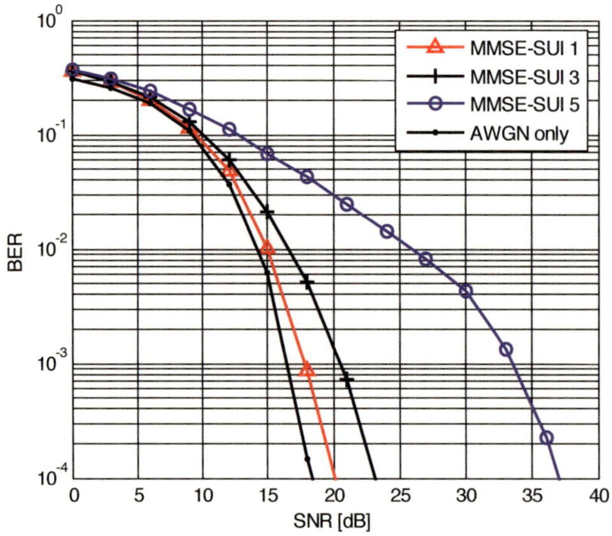

Figure 12. BER Performance comparison (16QAM, *N*=512).

4. WiMAX with compander for PAPR reduction

The system will be used in this work is shown in Figure 13. The transmitter section maps a random data bit sequence, into a sequence of QAM symbols. The QAM symbols are partitioned into N-length blocks and modulated onto the sub-carriers of an OFDM modulator via the inverse Fast Fourier Transform (IFFT). Before guard interval insertion, prior to transmission, the signal is companded with μ-law compander. The companded signal is then passed through the amplifier at the transmitter, which distorts the signal according to nonlinear solid state power amplifier models. After transmission, the signal is passed through SUI channels and corrupted with AWGN, compensation, μ-law expansion, and OFDM demodulation.

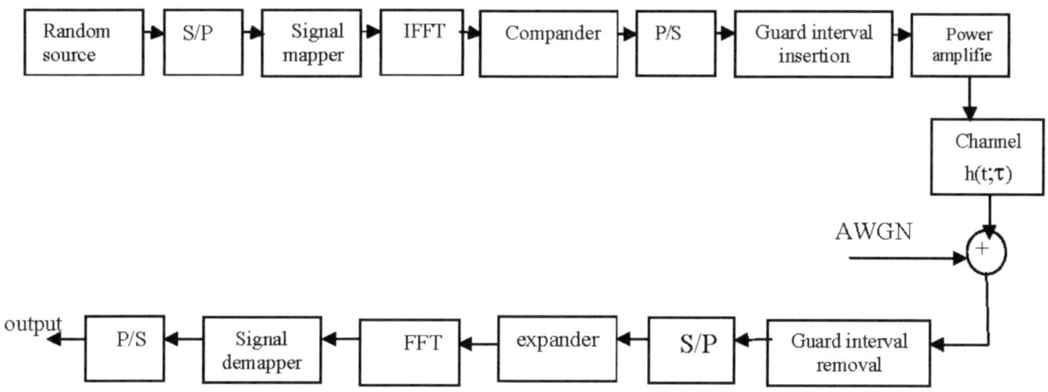

Figure 13. Simulation model of WiMAX system.

Compander (μ) was described in detail in chapter 3 where μ is the parameter controls the amount of compression.

5. Simulation results

The performance of proposed WiMAX is evaluated using computer simulation. In this study, the channel is assumed to be known perfectly at the receiver. The simulated system employs an OFDM signal with $N = 256$ subcarriers, among which *192* data carriers (*QPSK, 16QAM or 64QAM* signal mapping), *8* pilots, the others are nulls, guard interval $T_g = (1/4)T_b$, where T_b is the useful symbol time, and sampling frequency=9.12 *MHz*. For simplicity, uncoded OFDM will be employed. When applying the compander/expander, we must take into account some system constraints: Neither the pilots nor the guard intervals are allowed to be modified, for reasons of standard compliance.

5.1. CCDF performance

Figure 14 shows the performance improvement of the proposed system over a conventional system, i.e., without PAPR reduction, for the μ- law for different value of μ using 16 QAM signal mapping. At the probability of 10^{-3}, the PAPR is almost 1.1 dB, 2.55 dB, 4 dB and 4.5 dB smaller than conventional system, for μ=2, μ=4, μ=13 and μ=64, respectively. The same results are obtained when A-law is considered. Figure 15 shows the CCDF performance of the companding method at μ=13 compared with that of the system that is used clipping technique for reduction where CR = 2.

Figures 16 and 17 show the improvements which are obtained when 64 QAM and QPSK modulation are used, respectively.

5.2. BER performance

The BER performance of proposed WiMAX is evaluated here. In this study, the channel is assumed to be known perfectly at the receiver. The modulated signal is affected by SUI multipath channels and AWGN. Figure 18 shows the effect of different values of μ on the proposed system. In the following μ=13 is chosen to compromise between CCDF and BER performance. In Figure 19, at BER=10^{-4}, SNR over SUI 1 decreases by 8.2 dB better than the conventional system. The amount of improvement in SNR at BER=10^{-4} over SUI 3 and SUI 5 is 7.8 dB,and 5.2 dB better than the conventional system by using μ- law as shown in Figures 20 and 21, respectively.

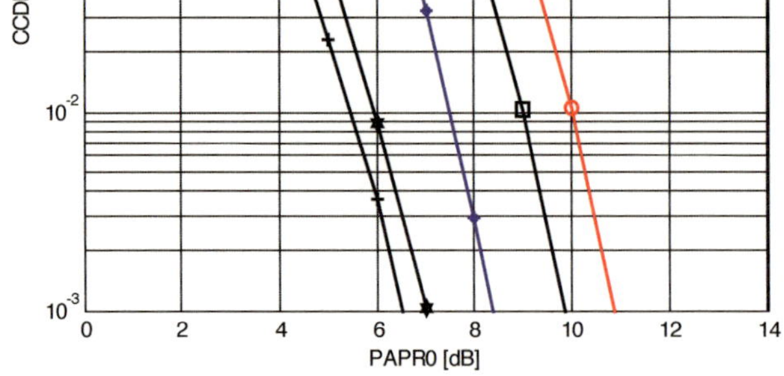

Figure 14. CCDF of PAPR with different μ (16-QAM constellation)

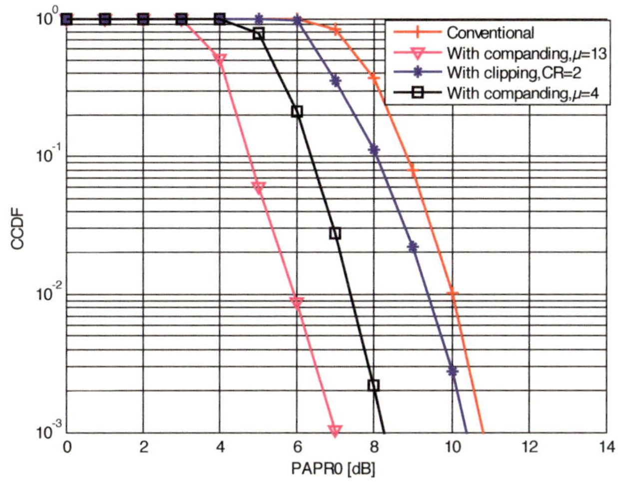

Figure 15. CCDF Performance comparisons with 16QAM.

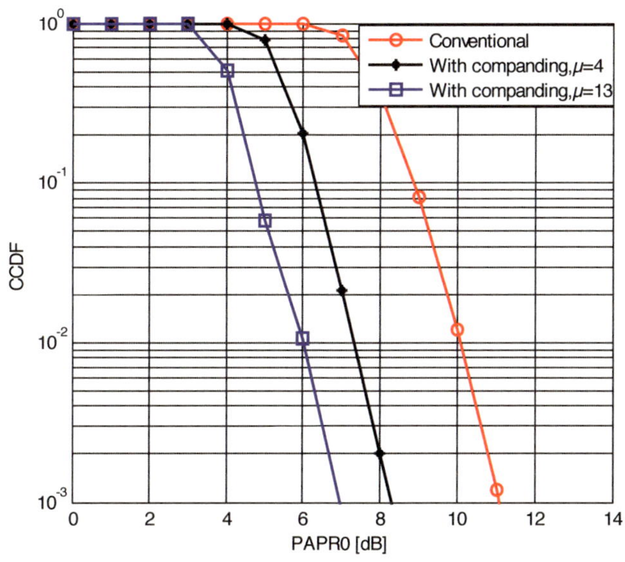

Figure 16. CCDF of PAPR with the proposed system (64QAM).

Figure 22 shows the system performance over AWGN only, where the amount of improvement in SNR is 10.7 dB at BER=10^{-4} than the conventional system. The BER performance of the companding method compared with the system that is used clipping technique where CR = 2 and CR=1.4 is shown in Figure 23.

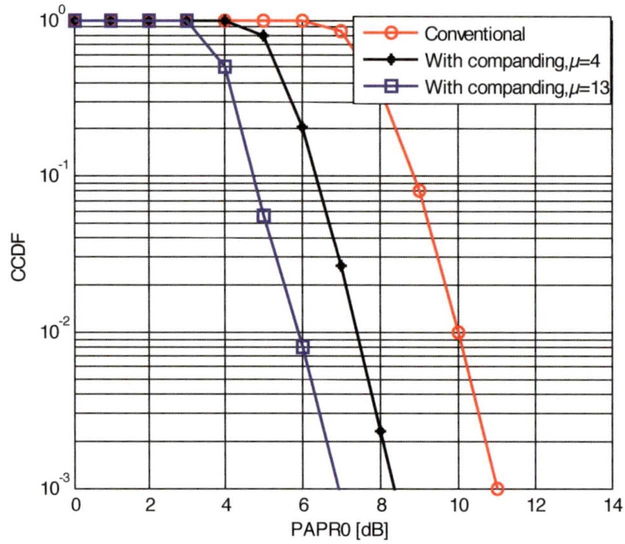

Figure 17. CCDF of PAPR with the proposed system (QPSK).

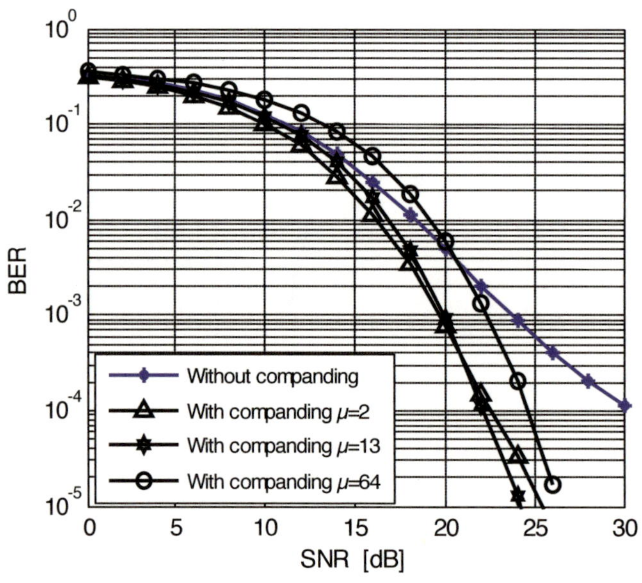

Figure 18. BER Performance with different μ.

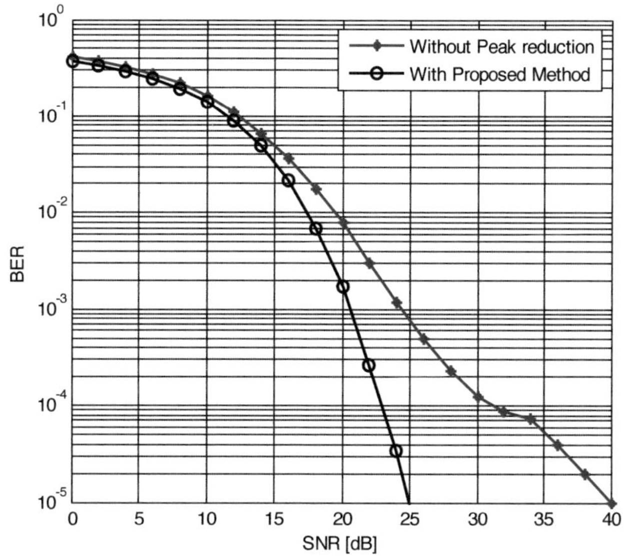

Figure 19. BER Performances over SUI 1 and AWGN.

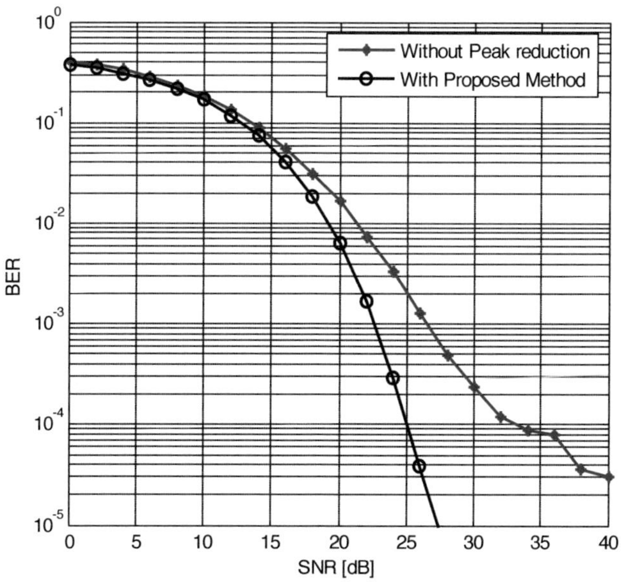

Figure 20. BER Performance over SUI 3 and AWGN.

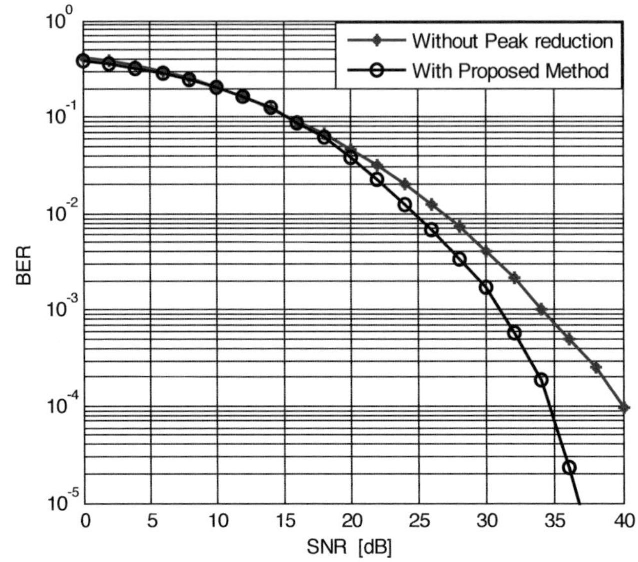

Figure 21. BER Performances over SUI 5 and AWGN.

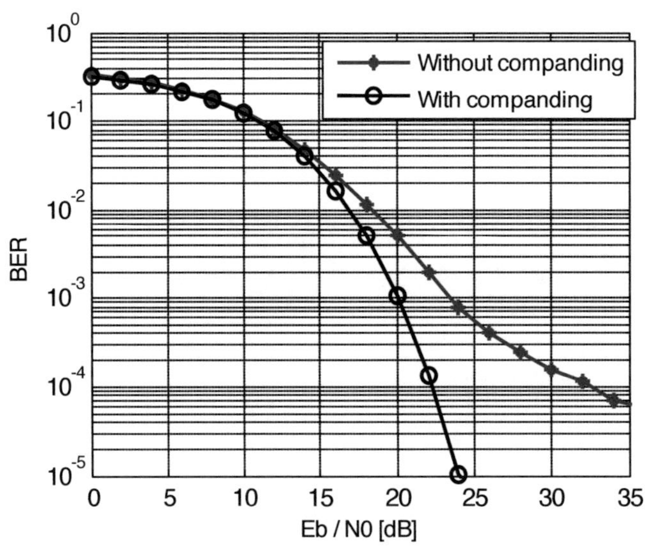

Figure 22. BER Performance over AWGN only.

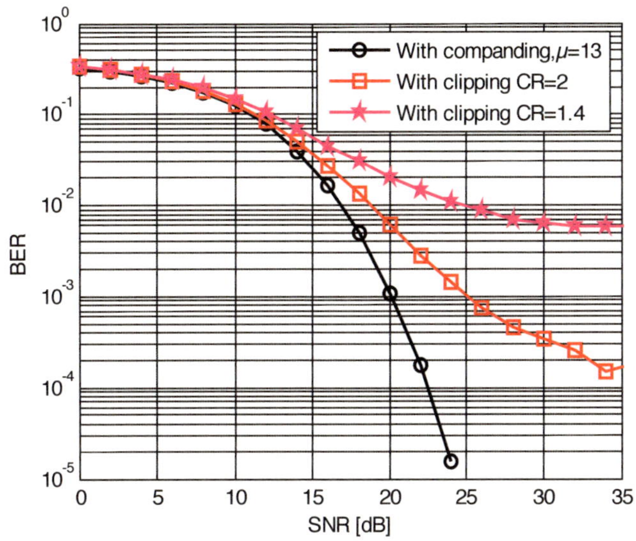

Figure 23. BER Performance comparisons.

6. Summary

We have investigated the PAPR in the WiMAX system. We investigates the simulation performance of WiMAX OFDM PHY Layer in the Presence of Nonlinear Power Amplifier and the new method is suggested to reduce PAPR where the simulation results show that, the peak power reduces by about 4 dB and SNR also decreases more than 5 dB at BER =10^{-4}. The Stanford University Interim (SUI) channel models are selected for the wireless channel in the simulation. Moreover the equalizer performance with SUI multipaths channel is explained.

Author details

Mona Shokair* and Hifzalla Sakran

*Address all correspondence to: shokair_1999@hotmail.com i_shokair@yahoo.com

Dept. of Electrical Communication, Faculty of Electronic Engineering, El-Menoufia University El-Menoufia, Egypt

References

[1] IEEE 802.16 and WiMAX. http://www.wimax industry.com/wp/papers/intel_80216_wimax.pdf (accessed 15 Decmber 2008).

[2] A Ghosh, D R Wolter, J G Andrews, and R Chen. Broadband Wireless Access with WiMax/802.16: Current Performance Benchmarks and Future Potential. IEEE Communications Magazine Feb. 2005; 43(2) 129-136.

[3] I Koffman, and V Roman. Broadband Wireless Access Solutions Based on OFDM Access in IEEE 802.16. IEEE Communications Magazine April. 2002; 40(4) 96-103.

[4] ETSI Broadband Radio Access Networks (BRAN); HIPERMAN; Physical (PHY) Layer. Standard TS 102 177, 2003.

[5] F H Gregorio. Analysis and Compensation of Nonlinear Power Amplifier Effects in Multi Antenna OFDM Systems. PhD thesis, Helsinki University of Technology, Finland; Nov. 2007.

[6] K Gorazd, J Tomaz, J Igor, and P Sreco. Effects of Nonlinear High Power Amplifier on the Area Covered by WiMAX Signal. IEEE Sarnoff Symposium : Conference Proceedings: March 2006.

[7] H Xiong, P Wang, and Z. Bu. An Efficient Peak-to-Average Power Ratio Reduction Algorithm for WiMAX System. Asia-Pacific Conference on Communications :Conference Proceedings: Aug. 2006.

[8] C Wei, Y Tianren, and W Hui. A New Method for Reduction of PAPR Using CORDIC Algorithm in WiMAX System. International Conference on Communications, Circuits and Systems : Conference Proceedings: 2006.

[9] S Hu, G Wu, J Ping, and S Li. HPA Nonlinearity Reduction by Joint Predistorter and Tone-Reservation with Null Subcarriers in WiMAX Systems. 4th IEEE International Conference on Circuits and Systems for Communications : Conference Proceedings: May 2008.

[10] S Haykin. Communication Systems. 4Th Edition John Wiley&Sons: In Tech; 2011.

[11] V Erceg. An Empirically Based Path Loss Model for Wireless Channels in Suburban Environments. IEEE JSAC July 1999; 17(7) 1205-1211.

[12] V Erceg, K V S Hari, M S Smith, and D S Baum. Channel Models for Fixed Wireless Applications. IEEE 802.16.3 Feb. 2001; Task Group Contributions.

[13] A Goldsmith. Wireless Communications. Cambridge University Press: In Tech; 2005.

[14] D Falconer, S L Ariyavisitakul, A B Seeyar, and B Edison. Frequency Domain Equalization for Single-Carrier Broadband Wireless Systems. IEEE Magazine of Communication April 2002; 40(4) 58-66.

http://dx.doi.org/10.5772/55380

[15] W Jeon, K Chang, and Y Cho. An Equalization Method for Orthogonal Frequency Division Multiplexing system in Time-Variant Multipath Channels. IEEE transactions on communications January 1999; 47(1).

[16] J Rinne, and M Renfors. An Equalization Method for Orthogonal Frequency Division Multiplexing System in Channels with Multipath Propagation Frequency Offset and Phase Noise. IEEE Global Telecommunications Conference : Conference Proceedings: 1996; 96(2) 1442–1446.

[17] H Sari, G Karam, and I Jeanclaudle. Frequency-Domain Equalization of Mobile Radio and Terrestrial Broadcast Channels. IEEE Global Telecommunications Conference : Conference Proceedings: 1994; 94(1) 1 – 5.

Chapter 3

A Mobile WiMAX Mesh Network with Routing Techniques and Quality of Service Mechanisms

Tássio Carvalho, José Jailton Júnior, Warley Valente,
Carlos Natalino, Renato Francês and
Kelvin Lopes Dias

Additional information is available at the end of the chapter

http://dx.doi.org/10.5772/55863

1. Introduction

The constant evolution of technologies for future wireless networks, along with the demand for new multimedia applications (voice, video,...) have led to the creation of new technologies for wireless communications. This is becoming one of the main challenges for this second decade of the third millennium, where new communications technologies must be sensitive to the need for bandwidth with high speed access, broadband in large coverage areas and the provision of services to an increasing number of users to ensure the next generation networks support for the content of new multimedia applications. Moreover, new technologies are an effective way of reducing physical barriers to the transmission of knowledge and transaction costs over fixed networks [1] [2]. Along with the creation of these wireless technologies, one of the current operating modes that is emerging is the mesh mode.

WMNs (Wireless Mesh Networks) are a special kind of MANET (Mobile Ad Hoc Network) and this research started out from the study and development of the MANETs. Compared with traditional networks, WMNs have many useful characteristics and peculiarities, such as dynamic self-organization, self-configuring, self-healing, high scalability and reliable services and are able to balance traffic and provide support to drop connections to fixed or mobile clients. In this way, it can prevent the decline of its services and avoid problems with flows where there is a need for bandwidth and high rates that are constantly required. This is achieved through a reconfiguration that always seeks the best alternative path to a better distribution of network traffic. Currently, many standard groups are improving the specifications of mesh networks from IEEE 802.11s to Wi-Fi (Wireless Fidelity), IEEE 802.15.4 to

Bluetooth and IEEE 802.16j to WiMAX (Worldwide Interoperability for Microwave Access) to multi-hop relay that will be the subject under study in this chapter.

The mobile WiMAX (Figure 1) is a technology based on IEEE 802.16 standard [3] developed as a feasible and attractive solution to these problems. It provides access to wireless broadband, especially an enabling context-sensitive network for the FI (Future Internet) with new multimedia applications, connectivity services for handover scenarios, long distances reaching the last mile, mobility management and mechanisms that improve communications with support for bandwidth and throughput metrics. These influence the network QoS (Quality of Service) with a certain level of end-to-end quality for multimedia applications through the management of layer 2 (Link Layer / MAC) and layer 3 (Network Layer / IP) for the provision of better services that give support to multimedia applications such as video stream and VoIP (Voice over Internet Protocol) that require real-time data delivery [4] [5].

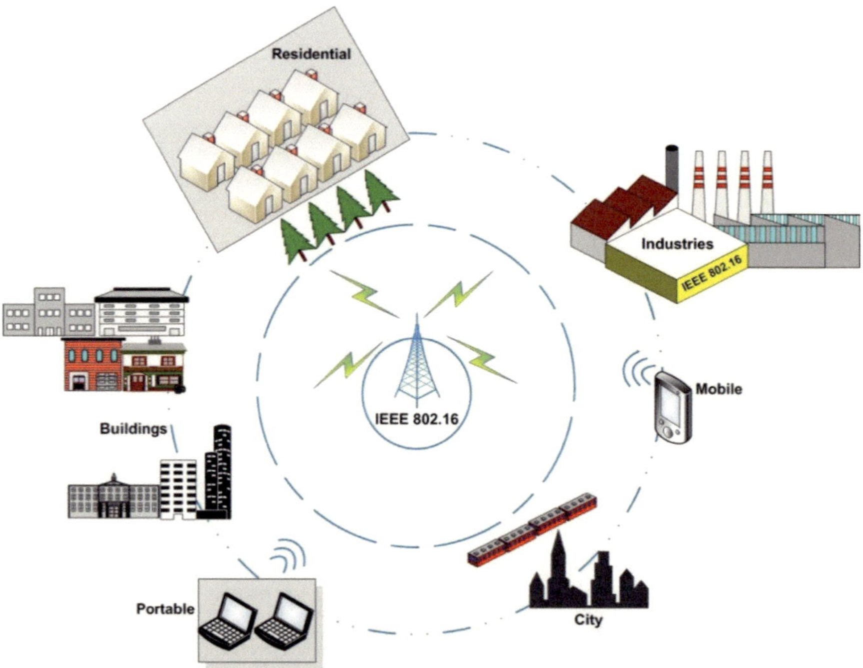

Figure 1. IEEE 802.16 / WiMAX network architecture

However, it is not clear enough how far the behavior of the WiMAX mesh network can support real-time services such as video streaming and VoIP, especially in mesh operation mode. Thus, this study provides an analysis of this question by analyzing network performance measurements through the properties of an IEEE 802.16 mesh network in several real-time applications. The chapter helps investigate the influence of routing protocols and the benefits of QoS to the network, as well as measurements for clients in a WiMAX wireless mesh environment, by showing their impact on flows and the final quality of multimedia applications.

QoS metrics, known as the rate of packet loss, delay and throughput, are generally used to measure the impact of multimedia streams on the level of quality of service, viewed from the perspective of the network, but do not reflect the user experience or the quality. As a result, these QoS parameters do not reflect subjective factors associated with human perception. In order to overcome the limitations of the existing schemes to guarantee QoS in networks with multimedia streaming that take account objective and subjective factors, the tests also address the impact of QoS and routing protocols on final quality through the QoE (Quality of Experience) concepts. This is carried out by addressing the user's perspective as the end-to-end quality of the video stream, by studying, evaluating and validating the results of QoS and QoE incidents on the routing metrics [6].

This chapter will provide an overview of the main challenges of the WiMAX mesh mode with a focus on routing protocols and the effect of quality of service mechanisms on scenarios with mobile clients. The chapter will describe the importance of mesh networks and how they can provide quality service and quality of experience for customers. It will also explain the impact of multimedia applications on this network and the importance of choosing the best route to ensure the network provides higher quality communications.

This section has provided a brief introduction to the main aims of this chapter. The second section will describe the mesh networks and explain their topology and operations. The third section will examine the QoS in WiMAX mesh networks. The fourth section will focus on routing protocols and draw attention to their main advantages and disadvantages. The fifth section will show the results of simulation tests obtained from analyzing the routing protocols with QoS and QoE. The sixth section explains the significance of the findings and conclusions, and this is followed by the seventh section with the main references.

2. WiMAX mesh network architecture

Wireless mesh operation mode is one of the most effective network branches among the emerging technologies. This network can connect multiple wireless access points (known as nodes) and form a mesh network, which is a network of connections that provides broad coverage and enables multiple paths and routes of communication. It is able to balance the traffic load and provide support for fault tolerance, so that if a node goes down, the network can self-configure and self-heal to find alternative routes of access [7].

WMNs can be seen as one type of MANETs [8]. An ad-hoc network (possibly mobile) is a set of network devices that want to communicate, but have no fixed infrastructure available and no pre-determined pattern of available communication links. The individual nodes of the network are responsible for a dynamic discovery of the other nodes that can communicate directly with them, i.e. what are their neighbors (forming a multi-hop network). Ad-hoc networks are chosen so that they can be used in situations where the infrastructure is not available or unreliable, or even in emergency situations. A mesh network is composed of multiple nodes / routers, which starts to behave like a single large network, enabling the client to connect to any of them. In this way it is possible to transmit messages from one node to another in different ways. Mesh type networks have the advantage of being low cost, easy to deploy and reasonably fault tolerant.

In another analogy, a wireless mesh network can be regarded as a set of antennas, which are spaced a certain distance from each other so that each covers a portion or area of a goal or region. A first antenna covers an area, the second antenna covers a continuous area after the first and so on, as if it were a tissue cell, or a spider web that interconnects various points and wireless clients. What is inside these cells and covers the span of the antennas, can take advantage of the network services, provided that the client has a wireless card with the interface technology.

Mesh networks are networks with a dynamic topology that show a variable and constant change with growth or decline, and consist of nodes whose communication at the physical level occurs through variants of the IEEE 802.11 and IEEE 802.16 standard, and whose routing is dynamic. The image below (Figure 2) shows an example of a mesh network. In mesh networks, the access point / base stations area is usually fixed.

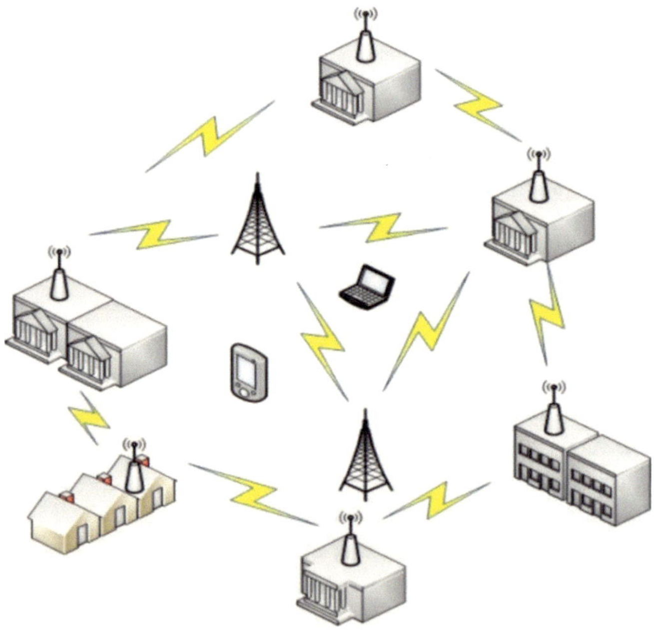

Figure 2. Mesh network

To achieve these goals, WiMAX networks can be structured into two operating modes: PMP (Point-to-Multipoint) and mesh networks, and the second is the focus of this chapter. Mesh mode is a type of operation that can interconnect multiple mobile clients together with many WiMAX base stations (nodes) and form a network of connections so as to provide a wide coverage area for mobile clients. All the clients can communicate with each other and there is no need for an intermediate node to act as the mediator of the network. In this mode, the IEEE

802.16 can provide broadband access with wireless support both single-hop and multi-hop settings [2].

The basic topology of an IEEE 802.16 mesh network consists of two participating entities, called Base Station (BS) and Subscriber Station (SS), displayed below (Figure 3). The BS is the central node, responsible for coordinating all the communication and providing connectivity to the client stations (fixed or mobile).

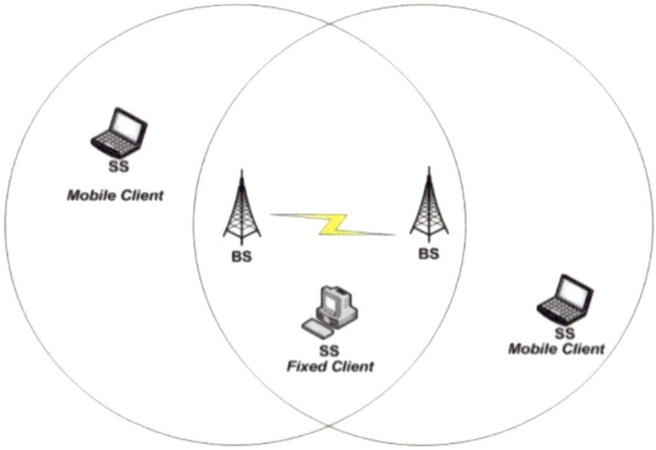

Figure 3. Basic topology of a WiMAX network network

Mesh networks reverse the idea of using a wired network to the backbone network and wireless access in the last mile. The backbone of a wireless mesh network comprises the router nodes that interconnect with the customers. As the nodes in the backbone network of this type have a fixed location and only the clients can be mobile, they may readily be fed, since they have no limiting power, and thus can rid themselves of many of the constraints of ad-hoc networks.

2.1. Operation

The most effective way to discover the operation of the mesh network is the routing protocol, which scans the different possible routes / paths of data flow, on the basis of a pivot table where devices such as BS select the most efficient route to follow to reach a goal, while taking into account that the greater the speed, the packet loss, or the faster the access to the Internet (and others). This scan is carried out several times per second and is transparent to the user, even when it occurs at re-routing access gateways, which are the nodes that have direct access to the internet.

An important feature of mesh networks is the concept of roaming, also known as a transparent handoff mobility scheme offering fast handoff in wireless networks. This makes it feasible for users to become mobile clients who can move around between network nodes without losing the connection at the time of exchange. The practical consequence is that the system allows geographical mobility. The system will always know which jumps are required for the request of a customer at any point in the network so that it can reach the Internet in the most efficient manner possible.

2.2. Challenges and problems

The growing interest in multimedia applications in mesh networks is accompanied by challenges that make the provision of QoS and group communication (multicasting) a more complex task. This complexity is the result, among other factors such as high mobility of the stations, which implies that there is a need to manage their locations and the environment and cope with the limitations of the devices involved, such as transmission quality in a wireless environment, bandwidth scarcity, etc.

Mesh networks have good prospects of being the solution to a series of problems in the provision of access services, since they are flexible, dynamic and potentially low cost [9]. However, for this to become effective there is much that needs to be improved and developed.

Besides routing, the major problems in mesh networks are scalability and security. The first can be defined as the level of acceptable service packages in the presence of a large number of nodes in the network. An important factor is the potential reduction in performance when there are an increased number of nodes. Hence, any protocol layers involved should be scalable. The security schemes proposed for ad hoc networks can be adopted for mesh networks, although most of these solutions have not been studied in depth and there are still problems that prevent them from providing authentication and reliability to clients.

Today the provision of QoS to any network is mandatory. When the mesh networks follow these steps, with the growth of multimedia applications, the services often seek a guaranteed bandwidth and QoS requirements, as a result of the growth of multimedia applications [2] [10]. In addition, they know that choosing the best path routing is an important decision for the WMNs to enable them to provide a wide range of services to different client types, each with their own peculiar characteristics. Provisioning QoS in mesh networks is not devoted to a single task layer. It requires the joint effort of all the layers, and specific strategies for signaling quality of service using resource reservation and QoS for the data link layer.

Owing to this and a number of other problems, when compared with other wireless network models, the mesh networks pose a special challenge, because the wireless environment is shared by adjacent nodes and the topology may change dynamically in the same way as the mobility of the nodes and input / output in the same network. As a result, QoS has become a key area of research of comparable importance to algorithms.

3. Quality of service in WiMAX networks

WiMAX has been developed with QoS in mind. Five different service classes have been introduced for different applications and packets from different service classes and are being handled on the basis of their QoS constraints. However, this mechanism can only be used in

http://dx.doi.org/10.5772/55863

the PMP (Point-to-Multipoint) mode. In the Mesh mode, QoS is maintained on a message-by-message basis.

In PMP mode, the WiMAX MAC layer uses a scheduling service to deliver and handle SDUs (Service Data Units) and MAC PDUs (Protocol Data Units) with different QoS requirements. A scheduling service uniquely determines the mechanism the network uses to allocate UL (UpLink) and DL (DownLink) transmission opportunities for the PDUs. WiMAX defines five scheduling services:

1. Unsolicited grant service (UGS): This is designed for the real-time constant bit rate (CBR) applications such as T1/E1 and VoIP. Unsolicited data grants are allocated to eliminate the overhead and latency of the request/grant process. During the connection establishment phase, maximum sustained traffic rate is declared and BS assigns fixed bandwidth grants in each frame accordingly.

2. Real-time polling service (rtPS): This is designed to support real-time services that generate variable-size data packets on a periodic basis, such as MPEG (Motion Pictures Experts Group) video. In this scheduling service, the BS provides unicast polling opportunities for the MS to request bandwidth. The unicast polling opportunities are frequent enough to ensure that latency requirements of real-time services are met.

3. Extended real-time polling service (ertPS): This scheduling service combines features from UGS and rtPS service classes. An initial ensured bandwidth allocation is carried out as in UGS and then this allocated bandwidth can be decreased or increased as in the case of rtPS.

4. Non-real-time polling service (nrtPS): This scheduling service is the most appropriate for the delay tolerant applications. As in rtPS, dedicated periodic slots are used for the bandwidth request opportunity, but with much longer periods. In nrtPS, it is allowable to have unicast polling opportunities, but the average duration between two such opportunities is in the order of a few seconds, which is large compared to rtPS. All the MSs belonging to the group can also request resources during the contention-based polling opportunity, which can often result in collisions and additional attempts.

5. Best effort (BE): This provides very little QoS support and is applicable only for services that do not have strict QoS requirements. It is for the traffic with no minimum level of service requirements. Like in nrtPS, contention slots are used for bandwidth request opportunities as long as there is space available [1] [2].

Classifiers are also present in the MAC layer of both the Base Station and Subscriber Station, whose goal is classify and map service flow into a particular connection for transmission between the MAC peers. The mapping process associates a data packet with a connection, which also creates a link with the service flow characteristics of this connection [11].

In this architecture there are schedulers in both the Base Station (BS) and Subscriber Station (SS), whose goal is to determine the burst profile and the transmission periods for each connection, while taking into account the QoS parameters associated with the service flow, the bandwidth requirements of the subscriber stations and the parameters for coding and modulation. Figure 4 illustrates the WiMAX QoS Architecture in PMP mode.

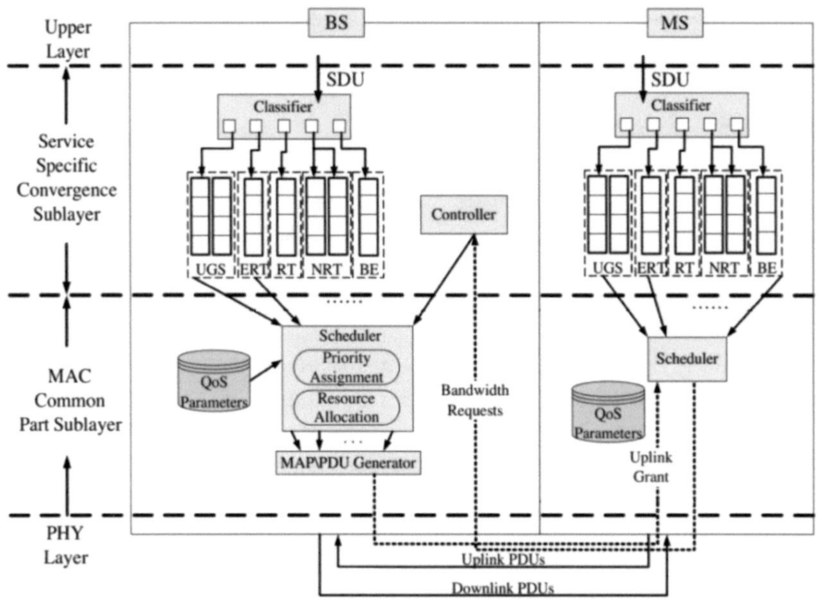

Figure 4. Architecture for IEEE 802.16 QoS

3.1. QoS in WiMAX networks in mesh mode

In a WiMAX mesh network, a "Mesh BS" (MBS – mesh base station) provides the external backhaul link. The backhaul links connect the WiMAX network to other communication networks. There may be multiple Mesh BSs in a network; other nodes are known as "Mesh SSs" (MSS – mesh subscriber stations). In point-to-multipoint mode, the SSs are under the direct control of the BS. In Mesh mode, the uplink and downlink is not clearly separated and SSs can communicate with each other without communicating with the BS.

3.1.1. IEEE 802.16 mesh frame

In the mesh mode, bidirectional links can be established between any of the WiMAX nodes, and the information is transmitted on a hop-by-hop basis. The system access follows a frame-based approach where each channel is divided in time into a series of frames. The number of frames in a series is defined during process of creating the network..

A frame is divided into two subframes: a control subframe and data subframe (Figure 5). The control subframes are used for carrying the information necessary for access control systems, bandwidth allocations, connection establishment and connection maintenance. The data subframes are used for carrying the packets of upper layers. The control subframe is divided into a number of transmission opportunities. The data subframe is similarly divided into a number of minislots.

There are two types of control subframes depending on their function. The first type of control subframe is the scheduling subframe in which nodes transmit scheduling messages. The second is the network configuration subframe in which nodes broadcast network configuration packets containing topology information, network provisioning information, and network management messages.

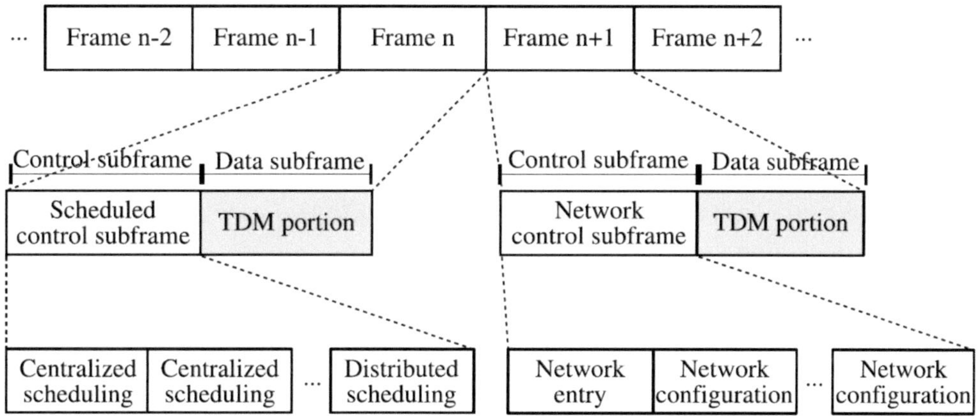

Figure 5. Mesh frame structure

The IEEE 802.16 mesh standard uses a combination of a 16-bit mesh node identifier (node ID) and a 16-bit connection identifier (CID) to identify the source and destination of every transmission. The CID in mesh mode is a combination of an 8-bit link ID and an 8-bit QoS description for the connection. All the communications occur in the context of a link, which is established between two nodes. One link will be used for all the data transmissions between two nodes. QoS is provisioned over links on a message-by-message basis. No services or QoS parameter are associated with a link, but each unicast message has service parameters in the header. Figure 6 shows the Mesh connection identifier (CID) construction which contains these service parameter fields.

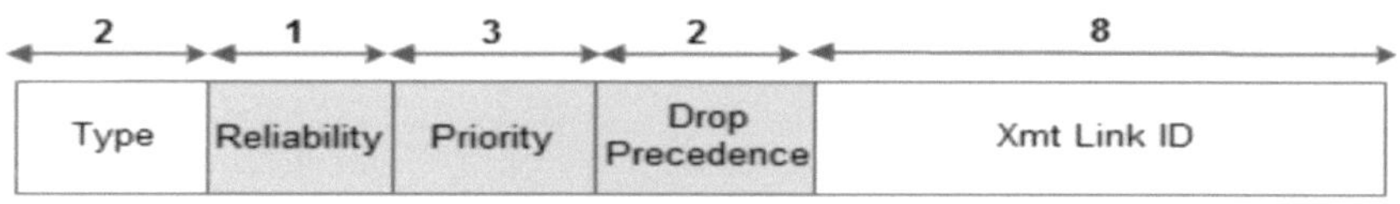

Figure 6. QoS bits in the mesh CID

The 8-bit QoS in the CID contains three definable fields: Reliability, Priority/Class, and Drop Precedence. Reliability refers to retransmit or not (0 indicates no retransmit while 1 indicates retransmit). Priority/Class refers to the priority of the packet. Drop Precedence refers to the probability of dropping the packet when congestion occurs [12] [13].

3.1.2. Default mesh QoS mechanism

In the mesh mode, a special MAC is defined in the IEEE 802.16, which provides two different types of scheduling mechanisms – centralized and distributed scheduling.

Centralized Scheduling (Mesh CS): the Mesh-BS is responsible for supplying resources for each link in response to resource requests. Mesh centralized scheduling messages transmitted in a scheduled control subframe are used for this purpose.

In centralized scheduling, when a node has packets to send to either other MSS or the MBS, it sends a request packet in the control subframe, using the Mesh Centralized Scheduling Message (MSH-CSCH message) to the MBS. The node sends one bandwidth request for each link it has and all requests belonging to that node are sent in one MSH-CSCH message. After receiving requests from all the MSSs in the network, the MBS applies its traffic scheduler to these requests, including its own traffic requests.

Based on the scheduler used in the MBS, these requests are granted, either wholly or partially. Then the MBS broadcasts these grants in a MSH-CSCH message. A grant packet describes the data subframe usage of a frame. This data subframe description belongs to a frame after the frame from which the grant is sent. Each MSS forwards this grant message to its children. However, these requests and grants only include the amount of data that a node can transmit [14]. Figure 7 illustrates how it works in mesh mode.

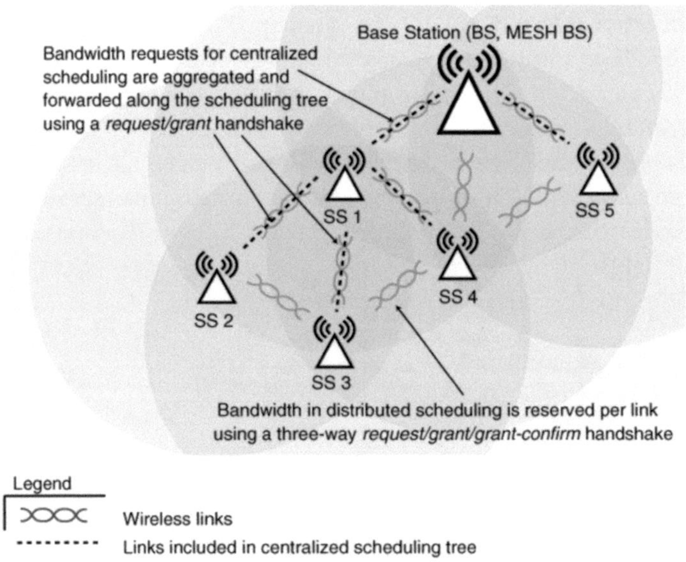

Figure 7. Overview of scheduling in the mesh mode

Distributed Scheduling (Mesh DS): The neighboring Mesh SS responds to a request with a corresponding grant for a link between two Mesh SSs. Mesh distributed scheduling messages are exchanged to perform this operation.

The scheduling policy for accessing data slots in coordinated distributed fashion, is not specified in the IEEE 802.16 standard. The standard only defines the Mesh Distributed Scheduling Message (MSH-DSCH message), and specifies the scheduling to avoid collisions between messages of different nodes. The MSH-DSCH message contains the scheduling information organized in Information Elements (IE): Request IE, Availability IE, Grant IE and Scheduling IE.

The scheduling procedure follows a three-way handshake to reserve the minislots. First, a node sends an MSH-DSCH message to one of its 1-hop neighbors, requesting a set of data slots. In the message, the node also includes the set of data slots that it has available for reservation. The 1-hop neighbor grants the request by replying with another MSH-DSCH message that specifies a set of data slots that confirms the availability of data slots at both nodes. Finally, the first node confirms the reservation of this set of data slots by repeating the grant in another MSH-DSCH message.

In contrast with point-to-multipoint WiMAX networks, the standard does not define scheduling services for Mesh WiMAX networks [13].

4. Routing protocols in wireless mesh networks

Currently, one of the main areas of mesh networks that is being studied, is the routing protocol used to find the best path to the base stations (or access points). This allows customers who use this type of technology to take advantage of their services in a more effective way and with efficient communication, as well as transferring their data stream through the wireless communication environment [15]. Routing is a service in which the router evaluates the possible paths to transmit packets to their destination, and determines the best route this packet should follow [16].

The concept of network performance optimization is carried out through the construction of the routing tree selection which is characterized by the topological properties that are independent when the network is being formed. The construction of the tree and arrangement of the nodes allows a distribution of the nodes that leads to a better chance of routing and optimization. The correlations between the topological parameters of the tree and the efficiency of the network must be estimated, and those that show the strongest correlations should allow the creation of the best trees and thus provide some routing and topology optimization [17].

Currently there are a number of routing protocols with several differences and similarities between them, that show the particular advantages and disadvantages when applied to mesh networks. Among these various routing protocols, there is no exists single protocol that can be claimed as the best. The reason for this is that they have several peculiarities and there not exists a protocol that is considered to be optimal for all scenarios. Each protocol has a unique characteristic, which makes it either suitable for a particular application.

By studying the scientific and academic papers in mesh networks, it is clear there has been a notable growth in the number of research studies in this area [18] [19]. There are currently several projects spread across the networks, some on a large scale. This is because of the benefits that can be derived from this mode operation, including the cost-effective deployment of broadband, and ease of access. Another potential element of fundamental importance is digital inclusion and the Future Internet which can provide services and comprehensive long-range topology wireless, suitable for specific topologies, with the implementation of QoS to meet the requirements of situations such as the next generation networks and the ever-increasing demand for multimedia applications and real-time.

As discussed earlier, mesh networks are a promising technology. However, to develop their full potential as a product, mesh networks require research in fields related to all the layers of the TCP / IP stack. Specifically in the routing area, there is a need for new protocols and critical metrics. However, the adoption of routing protocols of ad hoc networks in mesh networks, although possible, causes a number of problems and has drawbacks, such as the large number of control packets used for these protocols. The dynamics of an ad hoc network requires the constant assessment of the network topology, which is different from a mesh network with a static topology. Thus, a mesh routing protocol should be a more stable and less costly network.

However, before understanding routing protocol operations, it is necessary to understand the operation of routing algorithms that are of two kinds: non-adaptive algorithms (static) that calculate the route when the network is initialized and not based on a network topology and adaptive algorithms (dynamic) that take into account the topology and where to search for information.

Adaptive routing algorithms can in turn be classified in two ways:

a. Distance Vector (DV): Due to its applicability to packet routing on the Internet, this became known as Routing Information Protocol (RIP) or Distributed Bellman-Ford (DBF). This algorithm operates by enabling each router to maintain a table (i.e. a vector) which provides the smallest distance to each known destination and determines which line should be followed to get there. In a distance vector. routing is defined as a metric unit that will be the cost value of a path between nodes of a network. This metric unit could be the physical distance between nodes, the amount of hops (hops), the delay in transmission, the node congestion and other factors.

b. Link State (LS): This dynamic algorithm was devised with the purpose of solving the problem of distance vector routing, since it used the number of hops to the destination, although a packet could reach a destination by going a short way, ie with few hops. However, the link bandwidth could be small and the delay be greater. As a result,, the link state has arisen to find efficient routes, and is not concerned about the number of hops or the conditions under which the network is located.

4.1. Routing protocols

Among the ad-hoc networks, there are three basic types of routing protocols: proactive, reactive and hybrid [20] [21]. The proactive type requires us to maintain the route network for all possible destinations when there is a need to send a data packet. In reactive protocols, the nodes discover the destinations on demand. The hybrid protocols are those where there is only one set of nodes that provides periodically updated information on possible destinations.

4.1.1. Pro-active routing protocol (Table- driven)

This protocol requires all the network nodes to maintain routes to all possible destinations so that, when the need arises to send a data packet, the route that must be taken is known immediately. These protocols operate through their routing tables by exchanging messages continuously. Examples of proactive protocols are: : OLSR (Optimized Link State Routing Protocol), DSDV (Destination-Sequenced Distance-Vector) and WRP (Wireless Routing Protocol), the first, the OLSR is the representative of the protocols used for the following tests of this chapter.

The OLSR is a routing protocol developed for MANETs, and is an optimized link state protocol. The OLSR reduces the control packet size and the number of these packets that are sent to the network. This reduction in the number of control packets is achieved through the use of Multipoint Relays (MPR), which characterizes the OLSR. MPR is a node chosen from among the neighbors to send control packets, and the choice is made by the neighbors when there are only a hop of the node [22].

4.1.2. Reactive routing protocol (On-demand)

In the reactive protocols, the nodes discover the on-demand destinations, i.e. they do not require a route to the destinations where they have to send data, and seek the efficient use of resources like energy and bandwidth. Examples of reactive protocols are : AODV (Ad-Hoc On-Demand Distance Vector), DSR (Dynamic Source Routing) and TORA (Temporally Ordered Routing Algorithm). An examination of he AODV protocol, which is the representative of the reactive protocols used for testing, follows in this chapter.

The AODV routing protocol is a reactive protocol, i.e. the route to a destination node is discovered only when it wants to send a packet (data) to this node., This protocol enables dynamic routing, where the route of the packet can be changed in accordance with the route that the data is following, if the route used is unavailable. This discovery quickly results in new destinations [20] [23].

The AODV protocol is a protocol based on the Destination-Sequenced DistanceVector (DSDV) [19], and is created primarily to eliminate errors in DSDV, on account of the constant changes of topology and the large number of control messages between the network components. During the route discovery, the AODV protocol uses a traditional routing table as a storage mechanism. This only stores one entry, i.e. it only stores the next hop to the destination, unlike the DSR that stores multiple routes to the same destination and also stores the entire route from the source to a destination. The AODV is designed to be used in ad-hoc networks which have provided small numbers of nodes (up to thousands). The main purpose of the protocol is to adapt quickly and dynamically to the changing conditions of the network links, and find routes which can allow it to provide a desirable QoS. In this way it, avoids wasting bandwidth, minimizing memory usage and processing the nodes that act as routers.

4.1.3. Hybrid routing protocol

The hybrid is a protocol where a certain set of nodes, (only a limited number of nodes) periodically updates the information nodes / routes of possible destinations, and attempts to make a suitable use of the two previous approaches. Examples of hybrid protocols are : HWMP (Hybrid Wireless Mesh Protocol), ZRP (Zone Routing Protocol) and FSR (Fisheye State Routing); the HWMP protocol is the representative of the hybrid protocols used for the following tests of this chapter. HWMP is based on AODV [22] and also has an optional routing protocol, called RAOLSR (Radio Aware OLSR) based on OLSR [23] [24].

HWMP is a hybrid routing protocol. It has both re-active and proactive components. The creation of HWMP is an adaptation of AODV to radio-aware link metrics and MAC addresses. It is the basic, reactive component of HWMP. The on-demand path setup is achieved through the path discovery mechanism that is very similar to that of AODV. If a mesh point needs a path to the destination, it broadcasts a path request message (PREQ) into the mesh network. The hybrid routing protocols combine the best features.

4.2. Correlation between QoS and routing

QoS routing is an important parameter for the provision of guaranteed QoS in mesh networks. This issue has been exhaustively studied in wireless mesh networks. The aim of QoS routing for these networks is twofold: to find a best feasible path for each incoming connection in the presence of the underlying link interference and to optimize the usage of the network by balancing the load.

This chapter evaluates the routing problem in the IEEE 802.16 mesh networks. Unlike other routing strategies, this chapter is concerned with providing paths, mainly at certain QoS levels that guarantee traffic flows. The simulations will evaluate multimedia applications such as VOIP, video conference and other multimedia streams that have grown over the Internet, and verify the best qualifications between QoS and routing protocols by evaluating the major impacts on these two important factors in the WMNs.

The number of hops is the most common criterion that is adopted by traditional routing protocols. However, it is clear that these protocols are inadequate for multimedia applications, such as VoIP and video conferencing, which require QoS guarantees. Routing protocols with QoS, not only need to find the route with the shortest path, but the best route that meets the requirements of end-to-end QoS, regardless of the number of hops or how the routing protocols need to find the best routes through multiple hops. It, is necessary and important that the new protocols and routing algorithms also take into account the parameters and other measurements such as power consumption, the closeness of the backbone network output and

http://dx.doi.org/10.5772/55863

especially the quality over quantity link for users and the quality of wireless communications, while taking into account attenuation, signal quality and interference.

5. Evaluation and results

The Simulations experiments were carried out with the aid of Network Simulator version 2 [25] to show the performance of some routing protocols with QoS as network measure in WiMAX Mesh Network. For the WiMAX Mesh simulations it was used a module developed by the Network and Distributed System Laboratory [26] with extensions to use on PMP and mesh mode. The results compare four routing protocols: AODV, OLSR, HWMP Proactive and HWMP Reactive. Figure tal show the topology used for the tests, a random topology.

The simulation scenario chosen for the experiments were formed in a randomly generated with sixteen nodes, but that could easily represent a pre-existing base stations in a city, a rural area or a group of cities in proximity.The base stations act as routers through which network traffic will be routed through them choosing the best path according to its algorithms so that traffic is routed between source and destination.

The scenario (Figure 8) aims to test the choices of the best routes according to the algorithms / routing protocols and verify the flow and the delay due to these choices. The results are found in the simulations are evaluated along with the following analysis of these.

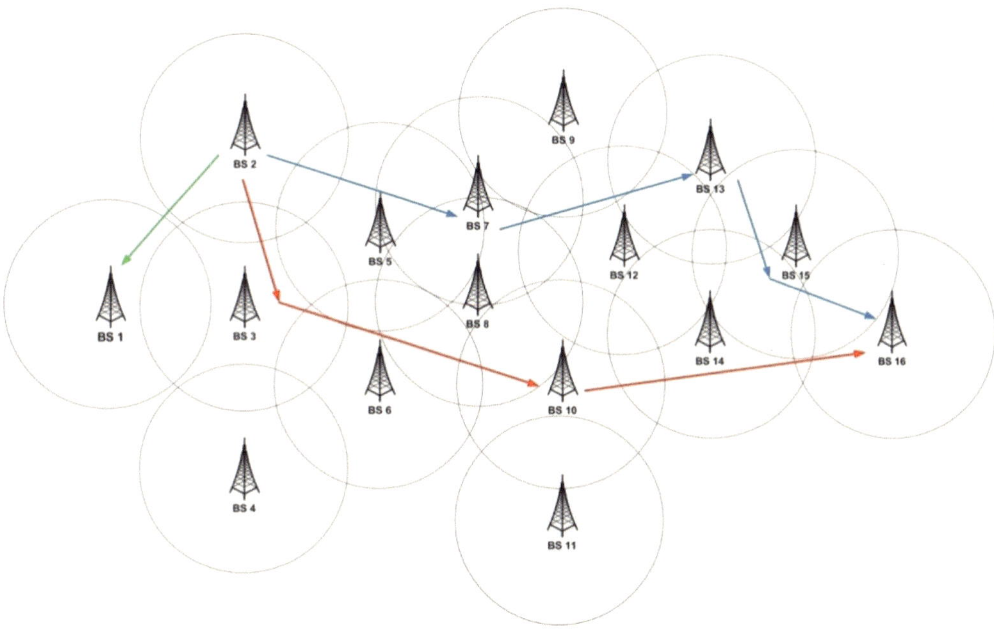

Figure 8. Simulated Topology

Faced with this scenario, routing protocols, based on their algorithms must choose the best route for that traffic out of the source node (node 2/BS 2) and reaches the destination node (node 16/BS16) and there is the question. What's the best route? The red route or blue route? Will would other routes? Perhaps green route. Certainly there are several routes and choosing each one behind certain characteristics and particular outcomes to the performance of this network and its communication. Simulated parameters presented below (Table 1).

Cover Area	1km
Frequency	3,5GHz
Standard	IEEE 802.16 (MESH)
Modulation	OFDM
Router WiMAX Mesh Number	16
Simulation Time	60s
Traffic	Video and CBR

Table 1. Simulated parameters

5.1. CBR traffic

In the first situation, the simulations were conducted with CBR traffic (1 MB), the transmission consists hop-by-hop by four routing protocols: AODV, OLSR, HWMP Proactive and HWMP Reactive. By the analyze of the throughput, achieved better performance result by HWMP Reactive (Figure 9). This result is because of the protocol in this scenario constantly keep checking the best route and always find a solution when faced with a new, always managing to optimize the flow through the best link at any given time.

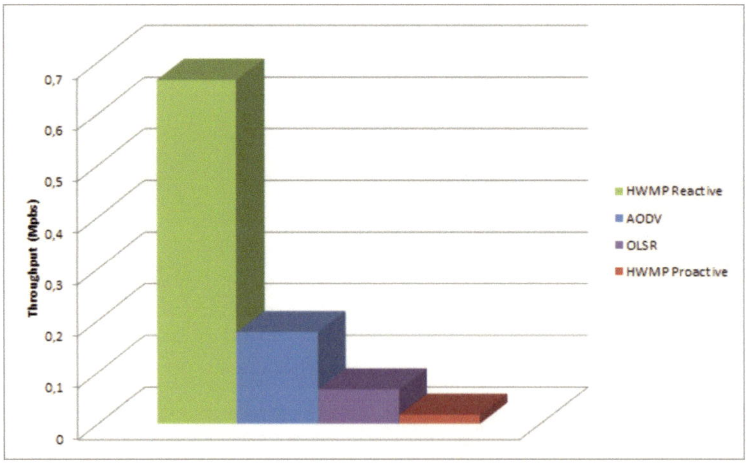

Figure 9. Comparison of CBR traffic throughput for the four routing protocols

http://dx.doi.org/10.5772/55863

The result of the hybrid routing protocol show the better results in comparison with other protocols presented here. In other protocols, it takes a long time to find a best route for data flow and sometimes, take congested routes, which reduces the throughput of the network.

5.2. Video on CBR traffic

In the second situation, the simulations were conducted with Video and CBR traffic (as background traffic). The transmission consists hop-by-hop by four routing protocol: AODV, HWMP Proactive, HWMP Reactive, OLSR. When we analyze the throughput, we observed a better performance by AODV. This case was carried out by using the Evalvid tool [27] that allows control of real video quality called "Grandma". The video simulations parameters presented below (Table 2).

In this particular case the transmitted traffic behind will focus on some decrease in the quality of connections that take the main traffic to the destination and make the hybrid routing algorithms are flawed when compared to non-hybrid and in this case, can best AODV results in selecting the best route and consequently better results regarding the flow, providing a certain QoS to the end customer and the quality of multimedia applications used. The AODV establishes the route more faster than other protocols, for this reason it had better throughput and better video performance.

Parameters	Value
Resolution	352 x 288
Frame Rate	30 Frame/sec
Color Scale	Y, U, V
Packet Length	1052
Packet Fragmentation	1024

Table 2. Video simulation parameters

Traditionally, the performance of network archictetures have been evaluated through Quality of Service (QoS) metrics. QoS is defined as the ability of the network to provide a service at an assured service level. QoS is also a commonly used metric set (e.g., throughput, packet loss, delay, jitter, handoff dropping and blocking probability) to represent the capability of a network to provide guarantees to selected network traffic. QoS considers parameters of a network that can be easily measured, but do not tell how the service is perceived by users. To satisfy the user-centric approaches, QoE is used to quantify the perception of the user about the quality of a particular service or network. The QoEmetrics confirm the previous statement.

The PSNR (Peak Signal to Noise Ratio) [6] [28] is the most traditional QoE/video metric, which estimates the video quality in decibels, comparing the original video with the video received by the user considering the aspects of luminosity. Figure 10 shows the better video quality using the PSNR statistics (Table 3).

For each PSNR range values, there is a qualification for the received video by the user. The Table II shows the PSNR range quality:

PSNR (dB)	Quality
> 37	Excellent
31 – 37	Good
25 – 31	Fair
20 – 25	Poor
< 20	Bad

Table 3. PSNR range

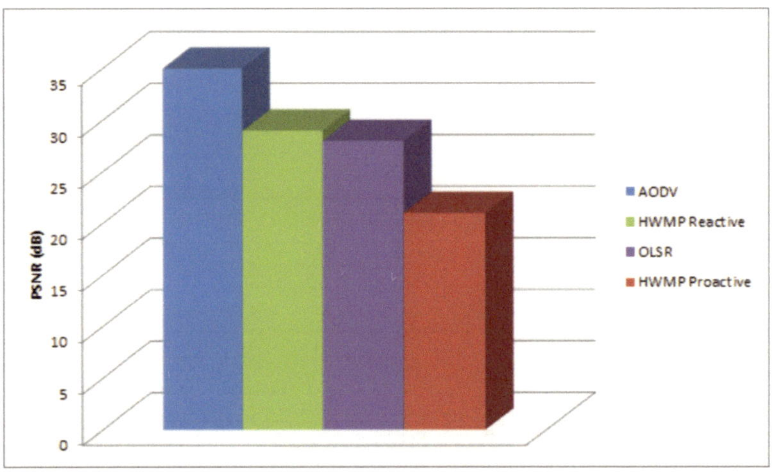

Figure 10. PSNR

The Structural Similarity (SSIM) [6] [28] metric evaluates the received video by the user taking into account the characteristics of the HVS (Human Visual System). The SSIM examines the color, light and structure similarity. The SSIM value is expressed by a number between 0 and 1, where 0 means zero correlation with the original image and 1 means the exact same video. As can be noted (Figure 11) by analyzing the QoS metrics, AODV has the best closeness in quality compared to the original video and HWMP Proactive worse.

http://dx.doi.org/10.5772/55863

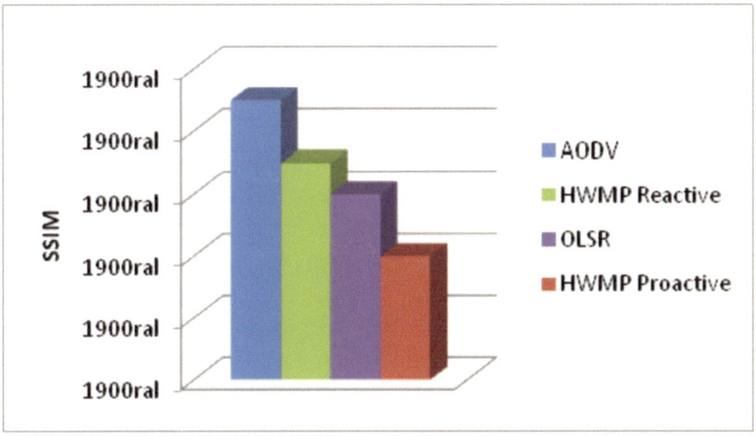

Figure 11. SSIM

The Video Quality Metrics (VQM) [6] [28] as MSU VQM metric also compares the original video with the video received by the user. They are considered the most complete metrics because compare the following aspects: noise, distortion and color. Again, AODV (Figure 12) has the best values because the smaller the value of this metric, better the video quality.

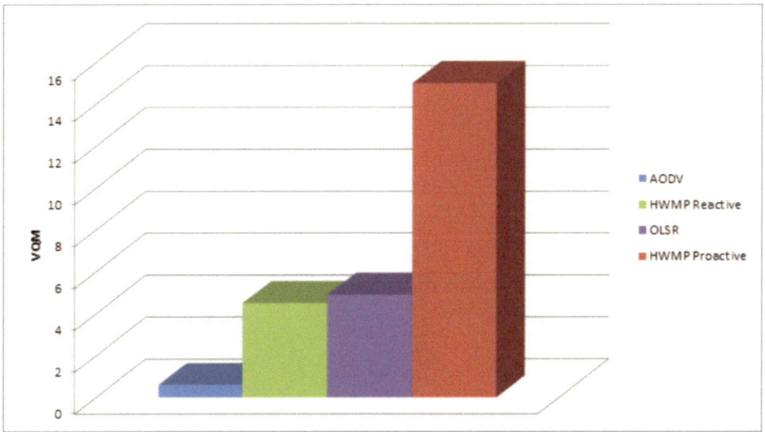

Figure 12. VQM

The evaluation of routing protocols are clear when we look at the frames in Figures. The frame number 100 was selected to compare the quality. As we can see, the AODV presents the best results, followed by hybrid reactive, OLSR and hybrid proactive protocols respectively (Figure 13).

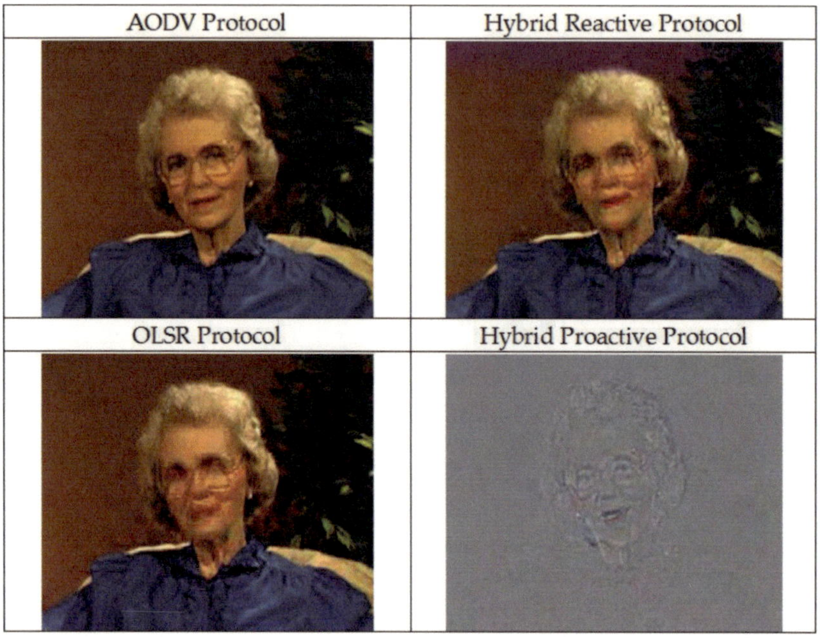

Figure 13. Result of frame 100 for the four routing protocols

The Figure 14 shows the delay average over time for the four routing protocols, showing their results in this network metric. AODV and HWMP reactive had the lowest delay.

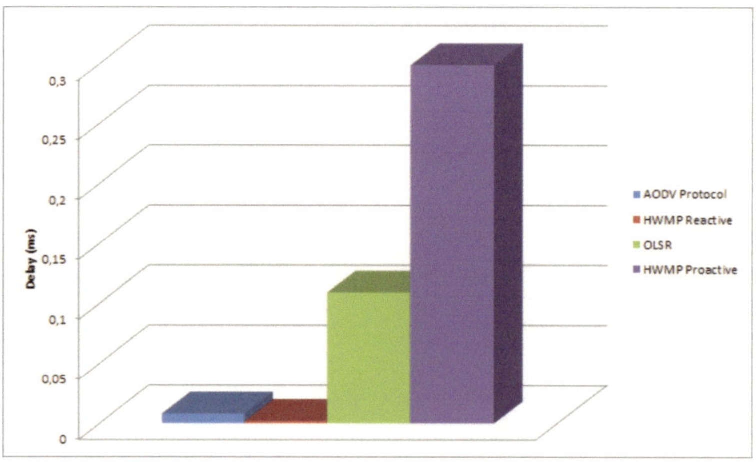

Figure 14. Delay average over time

If a routing protocol takes longer time to find the best route and took and thus decide to use your communication path, in normal situations, present a lower performance as measures of network for those with a behavior in choosing the fastest route.

6. Conclusions

This chapter showed an initial study on wireless mesh networks, pointing out the main goals of interest, challenges and issues encountered, highlighting the advantages and disadvantages of this mode of operation. During the work, focused on the IEEE 802.16 popularly known as WiMAX, a technology standardized by the WiMAX Forum as an alternative wireless communication with wide area coverage and bandwidth, providing high speed and mobility, important peculiarities in the context of next-generation networks in Future Internet.

The chapter provides a more detailed study of the WiMAX mesh mode, pointing to two very important points for this type of network the next generation of wireless communications: algorithms / routing protocols and QoS, mainly to meet the demands for new multimedia applications as VoIP, telemedicine, videoconferencing and other real-time applications that require a large bandwidth with constant to meet the constant flow needs, providing quality network metrics such as throughput and delay and qualitative results regarding the perception of the end user when evaluated on the QoE metrics, with valid results on human perception of quality in real end user.

The studies and validated through simulation showing what the main advantages of routing protocols when incidents of random scenario presented here, however, it is noteworthy that these results are specific to this scenario, not ensuring that the protocols achieve similar results in any type of scenario. It is important to mention that the protocols have different results as there is a variability of scenarios or data flow, increasing them or decreased them and so, some protocols may have better results in some scenarios and worse in others, there is certain variability.

Conclusively, routing protocols have advantages and disadvantages and present very particular results, and there is a protocol that presents the best results ever, nor how to choose the best route, nor as to the best results of QoS and QoE.

Simulation results shown that the AODV protocol provides the best results when analyzed on the scenario shown to video traffic, however, the hybrid routing protocol that operates in a reactive mode, gives good results and operates in hybrid form, could be better than AODV depending on some parameter variations. This makes us believe that the hybrid reactive would be a protocol that can be relatively good in all cases, although not an optimal model, can be efficiently and effectively providing a good alternative routes and relative quality to the end user about the prospect of QoS and QoE. The hybrid reactive and AODV protocol gave good results as the data flow rate and video quality, but could have different results in other settings and with other simulation parameters.

In some future work, the authors intend to make optimizations in routing protocols and mechanisms include your choice between a more complete analysis taking into account other important points beyond the amount of jumps as energy consumption and output communications to the Internet outside the backbone of the mesh network with algorithms that also take into consideration the proximity.

Author details

Tássio Carvalho[1], José Jailton Júnior[1], Warley Valente[1], Carlos Natalino[1], Renato Francês[1] and Kelvin Lopes Dias[2]

*Address all correspondence to: tassio@ufpa.br

1 Federal University of Pará, Brazil

2 Federal University of Pernambuco, Brazil

References

[1] J. G. Andrews, A. Ghosh, and R. Muhamed, "Fundamentals of WiMAX: Understanding Broadband Wireless Networking". Englewood Cliffs, NJ: Prentice-Hall, 2007.

[2] M. Kas, B. Yargicoglu, I. Korpeoglu, and E. Karasan, "A survey on scheduling in IEEE 802.16 mesh mode" Commun. Surveys Tuts., vol. 12, no. 2, pp. 205–221, 2010.

[3] IEEE 802.16e-2005 Part 16: Air Interface for Fixed and Mobile Broadband Wireless Access Systems Amendment 2: Physical and Medium Access Control Layers for Combined Fixed and Mobile Operation in Licensed Bands, 2005.

[4] M. Sollner, C. Gorg, K. Pentikousis, J. M. Cabero Lopez, M. Ponce de Leon, and P. Bertin, "Mobility scenarios for the Future Internet: The 4WARD approach" in Proc. 11th International Symposium on Wireless Personal Multimedia Communications (WPMC), Saariselkä, Finland, September, 2008.

[5] Y. A. Sekercioglu, M. Ivanovich, A. Yegin, "A Survey of MAC based QoS implementation for WiMAX networks", Computer Networks, 2009.

[6] Winkler, S. "Perceptual video quality metrics – a review, in Digital Video Image Quality and Perceptual Coding", eds. H. R. Wu, K. R. Rao, cha 5, CRC Press, 2005.

[7] Ian. F.Akyildiz, Xudong Wang. "A Survey on Wireless Mesh Networks", IEEE Radio Communications, September, 2005.

[8] Min Kim, Ilkyeun Ra, Jisang Yoo, Dongwook Kim, Hwasung Kim. "QoS Mesh Routing Protocol for IEEE 802.16 based Wireless Mesh Networks", Advanced Communication Technology, ICACT 2008. 10th International Conference. February, 2008.

[9] Grosh, S. Basu, K. and Das, S. K. "What a Mesh! An Architecture for Next-Generation Radio Access Networks", IEEE Network, October, 2005.

[10] Akyildiz, I. F., Wang, X. e Wang, W. "Wireless mesh networks: a survey". Computer Networks Journal (Elsevier), vol. 47, no. 4, p. 445-487. March, 2005.

[11] Y. A. Sekercioglu, M. Ivanovich, A. Yegin, "A Survey of MAC based QoS implementation for WiMAX networks", Computer Networks, 2009.

[12] Y. Li, Y. Yang, C. Cao and L. Zhou, "QoS Issues in IEEE 802.16 Mesh Networks", 1 st International Conference on Information Science and Engineering, 2009.

[13] E. O. Attia, A. S. Amin, and H. M. El Henawy, "Novel IEEE 802.16 Mesh Node Architecture to Achieve QoS in Coordinated Distributed Mode", 14th International Conference Advanced Communication Technology, 2012.

[14] M. Kuran, B. Yilmaz, F. Alagoz, and T. Tugcu, "Quality of Service in Mesh Mode IEEE 802.16 Networks," 14th International Conference on Software, Telecommunications and Computer Networks, 2006.

[15] Alireza Ghiamatyoun, Mohammad Nekoui, Said Nader Esfahani, Mehdi Soltan. "Efficient Routing Tree Construction Algorithms for Multi-Channel WiMax Networks", Computer Communications and Networks, 2007.

[16] Mahboubeh Afzali, Vahid Khatibi, Majid Harouni. "Connection Availability Analysis in the WiMAX Mesh Networks", Computer and Automation Engineering, 2010.

[17] Jerzy C. Nowiński, Piotr Gajowniczek. "Routing tree construction in WiMAX mesh networks". Telecommunications Network Strategy and Planning Symposium, 2010.

[18] Ruhani Ab Rahman, Murizah Kassim, Cik Ku Haroswati Che Ku Yahaya, Mariamah Ismail. "Performance Analysis of Routing Protocol in WiMAX Network". IEEE International Conference on System Engineering and Technology, 2011.

[19] Peng-Yong Kong, Jaya Shankar Pathmasuntharam, Haiguang Wang, Yu Ge, Chee-Wei Ang, Wen Su, Ming-Tuo Zhou and Hiroshi Harada "A Routing Protocol for WiMAX Based Maritime Wireless Mesh Networks". Vehicular Technology Conference, 2009.

[20] Hui Cheng and Jian Nongcao. "A Design Framework and Taxonomy Hybrid Routing Protocols in Mobile Ad-Hoc Networks". IEEE Communications Survey, 2008.

[21] Luo Junhai, Ye Danxia, Xue Liu, and Fan Mingyu. "A Survey of Multicast Routing Protocols for Mobile Ad-Hoc Networks". IEEE Communications Survey & Tutorial, 2009

[22] Clausen, T. and Jacquet, P. "Optimized Link State Routing Protocol (OLSR)", IETF RFC 3626, October, 2003.

[23] Perkins, Charles E., Belding-Hoyer, Elizabeth M., Das, Samir R. "Ad Hoc On-Demand Distance Vector (AODV) Routing", IETF RFC 3561, July, 2003.

[24] Nagham H. Saeed, Maysam F. Abbod, and Hamed S. Al-Raweshidy. "MANET Routing Protocols Taxonomy". IEEE Internacional Conference on Future Communications Networks, 2012.

[25] NS-2 The Network Simulator, "http://www.isi.edu/nsnam/ns/", 2012.

[26] NIST WiMAX, "http://ndsl.ie.cuhk.edu.hk/", 2012.

[27] Evalvid Tool-Set. "http://www.tkn.tu-berlin.de/research/evalvid/", 2012.

[28] Z. Wang, L. Lu, Bovik AC (2004) Video quality assessment based on Structural distortion measurement, Signal Processing: Image Communication, vol. 19, no. 2 2004.

Chapter 4

Hybrid AHP and TOPSIS Methods Based Cell Selection (HATCS) Scheme for Mobile WiMAX

Mohammed A. Ben-Mubarak,
Borhanuddin Mohd. Ali, Nor Kamariah Noordin,
Alyani Ismail and Chee Kyun Ng

Additional information is available at the end of the chapter

http://dx.doi.org/10.5772/55894

1. Introduction

In the past, the earlier cellular systems concentrated on voice calls as the main application that has to be considered to fulfil end-user requirements. However, nowadays with the variety of the user application and their requirements, the new 3G/4G systems have to consider many applications such as voice calls video streaming/conference, online gaming, peer-to-peer application and many other application and their different requirements as shown in Table 1 [1-2].

	Application	Type	Bandwidth	Delay
1	Multiplayer Interactive Gaming	Real-time	50-85 kbps	<100 ms
2	VoIP & Video Conference	Real-time	4-64 kbps (VoIP) 32-384 kbps (Video call)	<150 ms
3	Streaming Media	Real-time	5-128 kbps (music) 20-348 kbps (video clip) "/>2 Mbps (movie streaming)	<300 ms
4	Web Browsing & Instant Messaging	Non real-time	<250 kbps (instant messaging) "/>500 kbps (email/web browsing)	N/A
5	Media Content Downloads	Non real-time	"/>1 Mbps	N/A

Table 1. WiMAX Application Types

Although Mobile WiMAX promises to deliver the triple play services, handover mechanism still has some unsolved issues, which may affect the end user QoS requirements [3]. Cell selection is one of the main phases which may affect the user requirements after the handover process. After the Serving Base Station (SBS) advertisement message, MOB_NBR-ADV and the MS handover scanning, the MS will examine the collected PHY information on the neighbour BSs (nBSs) channel conditions to select the Target Base Station (TBS) for the coming handover [4]. The conventional cell selection scheme in Mobile WiMAX is based on signal quality, i.e. the nBS which has the best signal quality in terms of Received Signal Strength, will be considered as the TBS for the coming handover [5]. However, a single criterion like signal quality is not sufficient as a basis to choose the best BS for different end user application's requirements. As an illustration in Figure 1, suppose the MS is in an overlapping area of two or more BSs that have similar signal quality, there will be an ambiguity on which one will the MS choose for different user application requirements. Putting a cell selection criterion on signal quality entirely may make an MS choose a TBS with a good signal quality but one which may incur higher delay or smaller bandwidth, this may affect some real-time applications such as VoIP, video conferencing, streaming media, multiplayer interactive gaming.

Handover operation in Mobile WiMAX can be divided into two main phases; the pre-handover phase, and the actual handover phase. In the pre-handover phase the cell-reselection algorithm is one of the most important steps and this may affect the end-user QoS after the handover. On the other hand, the actual handover phase mainly focuses on handover execution and network re-entry [4]. Most of the published works are on the actual handover phase, about how to execute the scheme with a small number of messages [6-10], while there are only a few researches that focus on cell selection scheme.

The 802.16 standard [6] defines the receive signal strength indicator (RSSI) or carrier-to-interference plus noise ratio (CINR) as the handover trigger factor and cell selection scheme whereas WiMAX supports many multimedia data services therefore, it is not sufficient to let the signal strength be the only reference criteria.

In [11], the authors introduced an enhanced handover target cell selection algorithm for WiMAX network based on the effective capacity estimation and neighbour advertisement. The algorithm let each BS to estimate its idle capacity and broadcasts its effective idle capacity via the media access control (MAC) management message, MOB NBRADV, to help each MS to select the handover target cell. Therefore, the cell selection algorithm uses two criteria, idle capacity and signal strength as a weighted cost function to select the target cell. The result shows that this scheme enhances performance in terms of packet loss ratio and system throughput. Nevertheless, other criteria can be taken into account to provide better services for different user applications. In vertical handover schemes, i.e. handover between different network technologies, some researchers have introduced multi-criteria based network selection as described in [12-13].

Because of the increasing number of mobile stations (MS) and the support of high speed Internet and multimedia data services, the base station (BS) of WiMAX often works at a data rate close to maximum capacity. Therefore, it is not proper to let the signal strength to be the only reference criteria to choose the next cell. The MSs should decide intelligently to switch to another idle cell to get the best channel. In order to select a TBS that best meets the end-user application requirements this paper proposes a smart way for selecting the right TBS based on a set of multiple decision criteria. The method uses Analytic Hierarchy Process (AHP) method for criteria weighting and Technique for Order Preference by Similarity to Ideal Solution (TOPSIS) for the TBS ranking based on some criteria such as CINR, queue length, and bandwidth.

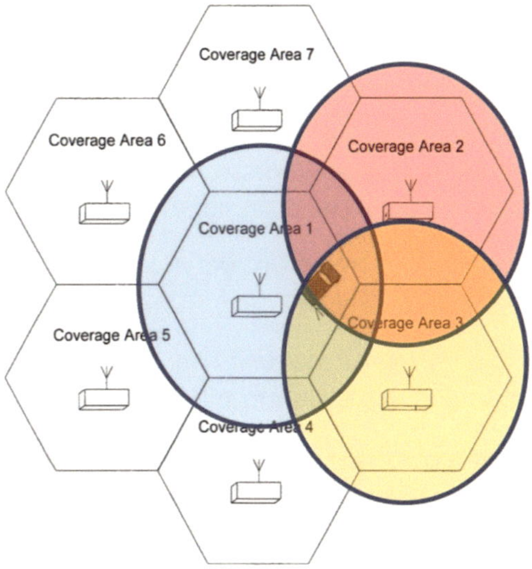

Figure 1. Cell selection

2. Handover cell selection criteria

The conventional scheme of cell selection in Mobile WiMAX is based on a single criterion which is signal quality. The nBS which has the best signal quality (ex. RSSI or CINR) will be considered as the TBS for the coming handover [5-6]. As highlighted earlier, this is inefficient because besides the signal quality each user's application has their own respective QoS requirements.

In this paper, two types of application are studied; VoIP as an example of real-time applications, and Media Content Downloads as an example of non real-time application [2]. As shown in Table 1, VoIP is a latency or delay sensitive application with low data rate demand. On the other hand, Media Content Download is a delay-tolerant application and generally demand high bandwidth. Thus, there are some other criteria that need to be considered when the cell selection decision is to be taken. In this paper, three criteria will be considered; they are CINR, BW, and congestion delay.

Carrier-to-Interference-plus-Noise-Ratio (CINR): The signal quality is the main criteria for choosing the TBS, but this is not a sufficient criterion.

Bandwidth (BW): This metric refers to the available bandwidth in WiMAX cell. It is simply the difference between the total capacities and the aggregated used BW in Kbps.

Congestion delay: This is the delay of packets due to queuing until they can be processed.

3. Multi-criteria decision making methods

The cell selection problem is about selecting one BS for handover among limited number of candidate BSs with respect to a set of different criteria. This is a typical Multiple Criteria Decision Making (MCDM) problem. In the study of decision making, terms such as MCDM is the problem of choosing an alternative solution from a set of alternatives, which are characterized in terms of their attributes [14]. The most popular classical MCDM methods are:

1. *WSM (Weighted Sum Model):* the overall score of a candidate BS is determined by the weighted sum of all the attribute values.
2. *TOPSIS (Technique for Order Preference by Similarity to Ideal Solution):* the chosen candidate BS is the one which is closest to ideal solution and the farthest from the worst case solution.
3. *AHP (Analytic Hierarchy Process):* decomposes the BS selection problem into several sub-problems and assigns a weight value for each sub-problem.

In this paper, we use TOPSIS as alternative score ranking based cell selection scheme and compare it with WSM, while AHP will be used for weighting the attributes or the criteria based on the importance of each criterion for the end users application.

Cell selection could be considered as an MCDM problem. For instance, suppose a user is currently connected to BS A1 and has to make a decision among two alternative candidate BSs: A2 and A3. Handover criteria considered here are CINR, BW, and congestion delay, which are denoted as: X1, X2 and X3, respectively. The decision problem can be modelled in a decision matrix Dm as shown in (1), where the capabilities of each candidate are presented.

$$Dm = \begin{array}{c} \\ A1 \\ A2 \\ A3 \end{array} \begin{array}{c} \begin{array}{ccc} x_1 & x_2 & x_3 \end{array} \\ \begin{bmatrix} x_{11} & x_{12} & x_{13} \\ x_{21} & x_{22} & x_{23} \\ x_{31} & x_{32} & x_{33} \end{bmatrix} \end{array} \tag{1}$$

Where x_{11} is the CINR, x_{12} is the BW and x_{13} is the congestion delay of BS A1. In similar way, x_{21}, $x_{22}$and x_{23} are the CINR, BW and congestion delay values of BS A2. In addition, x_{31}, x_{32} and x_{33} are the CINR, BW and congestion delay of BS A3.

Assume the user is using two types of applications; real-time application such as VoIP and non-real-time application such as Media Content Download. The traffic or application preference on handover criteria is modelled as weights assigned by the user on the criteria; for VoIP and Media Content Download which are shown in (2) and (3).

$$Wr = \begin{bmatrix} w_{r1} & w_{r2} & w_{r3} \end{bmatrix} \tag{2}$$

$$Wnr = \begin{bmatrix} w_{nr1} & w_{nr2} & w_{nr3} \end{bmatrix} \tag{3}$$

The CINR and congestion delay are considered as important for voice application because as studied by [15-16], to get a minimum jitter and delay the CINR has to be good enough, while the bandwidth and CINR are considered important for the non real-time application such as Media Content Download.

3.1. WSM (Weighted Sum Model)

WSM is the most popular multi criteria decision making (MCDM) method. It is the simplest way of evaluating the number of alternatives (m) in terms of a number of decision criteria (n) [17]. The overall score of an alternative is calculated as the weighted sum of all the attribute values as shown in equation (4).

$$A_i^{(WSM)} = \sum_{j}^{n} w_j x_{ij}, i = 1,2,3,.......m. \tag{4}$$

Where Ai is the evaluated score of an alternative, Wj is the weight value for criteria j, n number of criteria. Because the decision matrix value could be in different scales such as BW could be 10 Mbps and the cell load could be 50% or 0.50 the decision matrix has to have a comparable scale (normalized) by using (5) for the benefit criteria (i.e. stronger CINR, larger BW) and (6) for cost criteria (i.e. more delay). In (5) and (6) x_{ij} is the performance score of alternative A_i with respect to criterion x_j and r_{ij} is the normalization value of $x_{i\ j}$.

$$r_{ij} = x_{ij} / x_{jMAX}, i = 1,...m, j = 1,....n \tag{5}$$

$$r_{ij} = x_{jMIN} / x_{ij}, i = 1,...m, j = 1,....n \tag{6}$$

3.2. Technique for Order Preference by Similarity to Ideal Solution (TOPSIS)

TOPSIS is one of MCDM methods based on the concept that the chosen alternative should have the shortest distance from the positive ideal solution (PIS) and the farthest from the negative ideal solution (NIS) for solving a multiple criteria decision making problem. Briefly,

PIS is made up of all best values attainable criteria, whereas NIS is composed of all worst values incurred from criteria [14]. The calculation processes of this method are as follows. Normalize the decision matrix.

$$r_{ij} = x_{ij} / \sqrt{\sum_{i=1}^{m} x_{ij}}, i = 1,...,m, j = 1,....n \tag{7}$$

- Decision matrix is weighted using the weighting factor.
- Determine the ideal solutions A^+ and the negative-ideal solutions A^-.

$$A^+ = \left(\max_i v_{i\,j} \mid j \in J\right), \left(\min_i v_{i\,j} \mid j \in J'\right) \tag{8}$$

$$A^- = \left(\min_i v_{ij} \mid j \in J\right), \left(\max_i v_{ij} \mid j \in J'\right) \tag{9}$$

Where $v_{i\,j}$ is the weighted and normalized of the x_{ij}, while J is associated with the benefit criteria and J' is associated with the cost criteria.

- Calculate the separation of each alternative from the ideal solution, and the negative ideal solution.

$$\begin{aligned} S_{i+} &= \sqrt{\sum_{j=1}^{m} (a_{ij} - a_{j+})^2}, i = 1,...,n \\ S_{i-} &= \sqrt{\sum_{j=1}^{m} (a_{ij} - a_{j-})^2}, i = 1,...,n \end{aligned} \tag{10}$$

- Relative closeness to the ideal solution is calculated.

$$C_{i+} = S_{i-} / (S_{i-} + S_{i+}), i = 1,...,n \tag{11}$$

3.3. Analytic Hierarchy Process (AHP)

The Analytic Hierarchy Process (AHP) is a structured technique for organizing and analyzing complex decision problems. It has a variety of applications used around the world in different fields of decision situations such as business, industry, healthcare, engineering and education.

It provides a comprehensive and rational framework for structuring a decision problem, for weighting the decision criteria, and for evaluating alternative solutions [14, 18]. In this paper, we use the AHP method for weighting the decision criteria. The calculation processes of this method are as follows and shown in Figure 2.

The AHP method main steps are as follows:

1. Define the decision criteria and decompose the decision problem into different levels of the hierarchy.

2. Compare each factor to all the other factors within the same level through pairwise comparison matrix. The judgments in the AHP are made in pairs a_{ij}, relating the importance of criterion i to that of criterion j. The criteria are compared pair-wise with respect to the goal. A is a matrix indicating the importance of criterion i relative to criterion j as shown in Equation (12). Noticeably, the a_{ij}= 1 when i=j, while a_{ji}= 1/a_{ij}, which reflects the reciprocal importance of criterion j relative to criterion i.

$$A = \begin{bmatrix} a_{11} & a_{12} & a_{13} & \cdots & a_{1j} \\ a_{21} & a_{22} & a_{23} & \cdots & a_{2j} \\ . & & & & \\ a_{i1} & a_{i2} & a_{i3} & \cdots & a_{ij} \end{bmatrix} \tag{12}$$

1. After creating matrix A of comparison, the next step is to determine the weights of the criteria, in which w_i is the weight of objective i in the weight vector $w = [w_1, w_2, \ldots, w_n]$ for n criteria. To get this, the eigenvector is used.

The eigenvector can be calculated using the following steps:

1. Raise the A matrix to powers square.

2. Sum each column of the squared matrix.

$$a'_i = \sum_{j=1}^{n} a_{ij} \tag{13}$$

1. Then we divide each element of the matrix with the sum of its column, we have normalized relative weight. The sum of each column is 1 as shown in (14).

$$a''_{ij} = \frac{a_{ij}}{a'_i} \tag{14}$$

2. Check the Consistency Index (CI). The CI can be computed as the difference between the current and previous computed eigenvectors.

3. Previous steps are repeated until we get a very small value of CI.

4. Determine the weights of the criteria

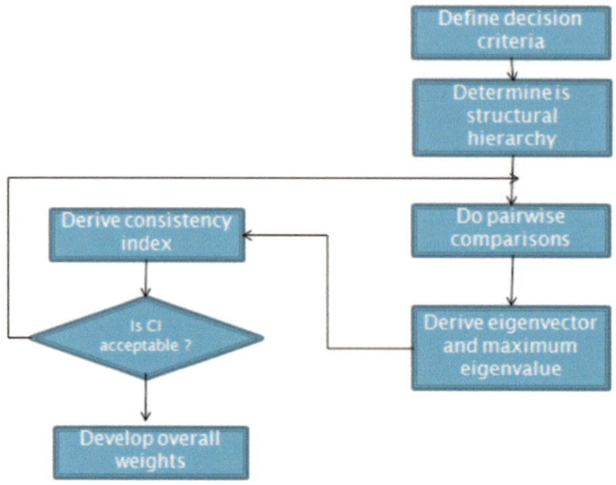

Figure 2. AHP calculation processes

4. Hybrid AHP and TOPSIS based cell selection

The conventional cell selection scheme in Mobile WiMAX is only based on the received signal strength. However, it is not sufficient to meet the different user application requirements that WiMAX promises to meet. The proposed scheme considers multiple criteria; CINR, BW and congestion delay for cell selection using some MCDM methods, AHP and TOPSIS. As shown in Figure 3, the proposed cell selection scheme can be divided into three main functions: "collecting info" which collects the decision criteria and network conditions, "criteria weighting" which processes criteria weighting using AHP methods based on application QoS requirements, and "alternatives ranking" which finalizes the process of cell selection using TOPSIS method.

4.1. Collecting Information stage

In this block the MS with BS assistance try to collect the information about some criteria that will help the cell selection decision. As highlighted earlier, this paper proposes three criteria decision, viz; CINR, BW and Congestion delay.

For the CINR criteria, the SBS regularly broadcast information about the nBSs through the neighbour advertisement messages (MOB_NBR-ADV). Additionally, the MS and the SBS can retrieve more information about the nBSs after the MS finished scanning, such as CINR and RSSI. The CINR is defined as the power ratio between a carrier and the interference and noise power [5].

$$CINR = 10\log_{10}\frac{C_r}{P_I + P_N} \tag{15}$$

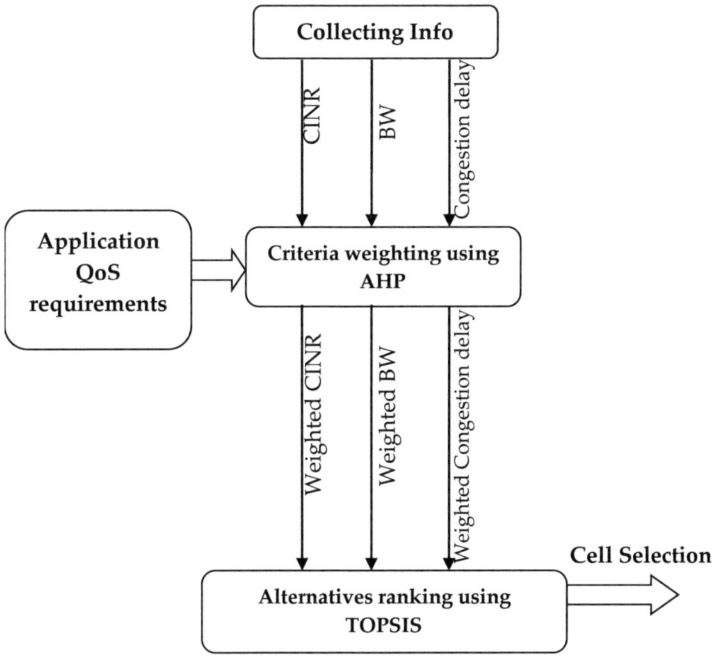

Figure 3. The proposed cell selection scheme

Where, C_r is the received carrier power, P_I is the interference power and P_N is the noise power.

The available BW in WiMAX cell metric can be estimated based on [11]; after scanning process, the MS can exploit the information element (IE) in the DL-MAP/UL-MAP messages and aggregate the number of the allocated downlink and uplink physical slots (PSs). The available downlink BW can be estimated as follows;

$$BW_{DL}(t) = \frac{1}{\tau} \quad S_{DL-free}(t-\tau,t) \quad C_{DL-slot}(t) \tag{16}$$

Where τ is the period of time that we can get number of frame, $S_{DL\ -free}$ is the number of unused PSs within τ while the $C_{DL\ -slot}$ is the number of bits that can be transmitted in one downlink PS.

Every nBS will inform the SBS its congestion delay through the backbone, and the SBS will send these values to the MS by the advertisement messages (MOB_NBR-ADV).

4.2. Weighting criteria using AHP

AHP is used to determine the weights for the three criteria -- CINR, BW and delay. It was reported that classical MADM methods cannot efficiently handle a decision problem with imprecise data that decision criteria could contain [14].

For that, the use of fuzzy logic can be used to deal with imprecise information and combine and evaluate multiple criteria simultaneously. Hence, fuzzy logic concept provides a robust mathematical framework in which handover decision can be formulated as a Fuzzy MADM.

The scale used is represented by the intensities between each other according to the fundamental scale. The fundamental scale is validated according to effectiveness and theoretical justifications according to [18]. The scale consists of nine levels. To make it even easier to judge, one can use a more restricted scale with five levels: 1 is equal importance, 3 is moderate importance, 5 is strong importance, 7 is very strong or demonstrated importance and 9 is extreme importance. Because different application have different requirements, in this paper, we consider two types of application -- VoIP as a real-time application example and Media Content Downloads as a non-real-time application example. Tables 2 and 3 show the pair-wise relative importance of the decision criteria for the real-time and non real-time applications respectively.

Real-time	CINR	BW	Cong. delay
CINR	1/1	5/1	3/1
BW	1/5	1/1	1/7
Cong. delay	1/3	7/1	1/1

Table 2. Pair-wise Matrix for real-time application

Non Real-time	CINR	BW	Cong. delay
CINR	1/1	3/1	5/1
BW	1/3	1/1	7/1
Cong. delay	1/5	1/7	1/1

Table 3. Pair-wise Matrix for non real-time application

Based on the AHP procedure explained earlier the weighting vector for the real time application and non real-time applications are as shown in equations (17), and (18).

$$w_r = [0.6186 \quad 0.0662 \quad 0.3149] \tag{17}$$

$$w_{nr} = [0.6186 \quad 0.3149 \quad 0.0662] \tag{18}$$

4.3. Alternative ranking using TOPSIS

After weighting the decision criteria using the AHP method, the TOPSIS method will be used to rank the alteratives available for the Target Base Station (TBS), and then choose the highest alterative score is chosen for the TBS. As mentioned earlier, TOPSIS concept is that the chosen alternative should have the shortest distance from the positive ideal solution (PIS) and the farthest from the negative. More explanation about this will be discussed in details in the numerical analysis section.

5. Numerical analysis

In this section a numerical analysis of the proposed scheme is performed, and then compared with the Weighted Sum Method (WSM) and conventional signal strength based cell selection. The criteria weighting based on user application requirements have been calculated previously, as shown in equation (17) and (18) for real-time and non real-time applications respectively.

Suppose that we have three candidate TBSs as shown in Figure 1. After the information collection stage, the MS will maintain the criteria values in a decision matrix, *Dm*(*A*). The decision matrix consists of three criteria, viz CINR in dB, BW in Kbps and congestion delay in ms for three candidate BSs as shown in (19).

$$\begin{array}{c} \qquad\qquad\quad CINR \;\; BW\,Cong \;\; Delay \\ Dm(A) = \begin{array}{c} BS1 \\ BS2 \\ BS3 \end{array}\begin{bmatrix} 23 & 2048 & 300 \\ 22 & 380 & 100 \\ 24 & 148 & 110 \end{bmatrix} \end{array} \tag{19}$$

Using the TOPSIS method, *Dm*(*A*) first has to be normalized first using equation (7), because the criteria use different units. The normalized matrix is shown in (20).

$$Dm(A') = \begin{array}{c} BS1 \\ BS2 \\ BS3 \end{array}\begin{bmatrix} 0.5770 & 0.9807 & 0.8960 \\ 0.5519 & 0.1820 & 0.2987 \\ 0.6021 & 0.0709 & 0.3285 \end{bmatrix} \tag{20}$$

After getting the normalized decision matrix, it is weighted using by multiplying it with the weighting factor. Two types of application have been considered, therefore, we have two weighting factors, which will produce two weighted decision matrices.

For the real-time application, the weighted decision matrix is as shown in equation (21).

$$Dm(A')_r = \begin{array}{c} BS1 \\ BS2 \\ BS3 \end{array}\begin{bmatrix} 0.3569 & 0.0649 & 0.2822 \\ 0.3414 & 0.0120 & 0.0941 \\ 0.3724 & 0.0047 & 0.1035 \end{bmatrix} \tag{21}$$

For the non real-time application, the weighted decision matrix is as shown in equation (22).

$$Dm(A')_{nr} = \begin{matrix} BS1 \\ BS2 \\ BS3 \end{matrix} \begin{bmatrix} 0.3569 & 0.3088 & 0.0593 \\ 0.3414 & 0.0573 & 0.0198 \\ 0.3724 & 0.0223 & 0.0217 \end{bmatrix} \tag{22}$$

To calculate the distance from the positive ideal solution (PIS) and the farthest from the negative ideal solution (NIS), equations (8) - (11) are used. CINR and BW are considered as benefit criteria and congestion delay is considered as a cost criteria. The ranking of the alternative TBSs for real-time and non real-time application are shown in equation (23) and (24) for real time and non real time application respectively.

$$\begin{matrix} & BS1 & BS2 & BS3 \\ R_r = [& 0.2479 & 0.7543 & 0.7484] \end{matrix} \tag{23}$$

$$\begin{matrix} & BS1 & BS2 & BS3 \\ R_{nr} = [& 0.8710 & 0.1724 & 0.1454\ ] \end{matrix} \tag{24}$$

Table 4 shows the summary of the ranking of the available alternative TBSs using TOPSIS compared with WSM and also illustrated in Figures 4 and 5 respectively.

MCDM methods	WSM			TOPSIS		
	BS1	BS2	BS3	BS1	BS2	BS3
VoIP(real-time)	[0.7640, 0.8942, 0.9097]			[0.2479, 0.7543, 0.7484]		
Media Content Download (non real-time)	[0.9298, 0.6917, 0.7015]			[0.8710, 0.1724, 0.1454]		

Table 4. Ranking of the available alternatives using WSM and TOPSIS

Using other wayS of viewing the cell selection results, Table 5 compare between TOPSIS, WSM and conventional signal strength-based cell selection.

	Signal strength	WSM	TOPSIS
VoIP	BS3	BS3	BS2
Media Content Download	BS3	BS1	BS1

Table 5. Selected cell using TOPSIS, WSM and the conventional signal strength schemes

For real-time applications, WSM ranks B3 as the best, and TOPSIS ranks B2 as the best. Both two results are reasonable, because all of them have good CINR and small delay. However, TOPSIS ranks B2 as the best, because it provides the lowest delay. So the TOPSIS looks like more sensitive to the criteria weighting. The conventional signal strength scheme ranks B3 as the best, because it provides the best CINR.

For the non real-time application, both WSM and TOPSIS ranks B1 as the best, because it has the highest BW while the conventional signal strength scheme ranks B3 as the best; although it has the best CINR, It provides the worst BW, which is an important criteria in the Media Content Download application.

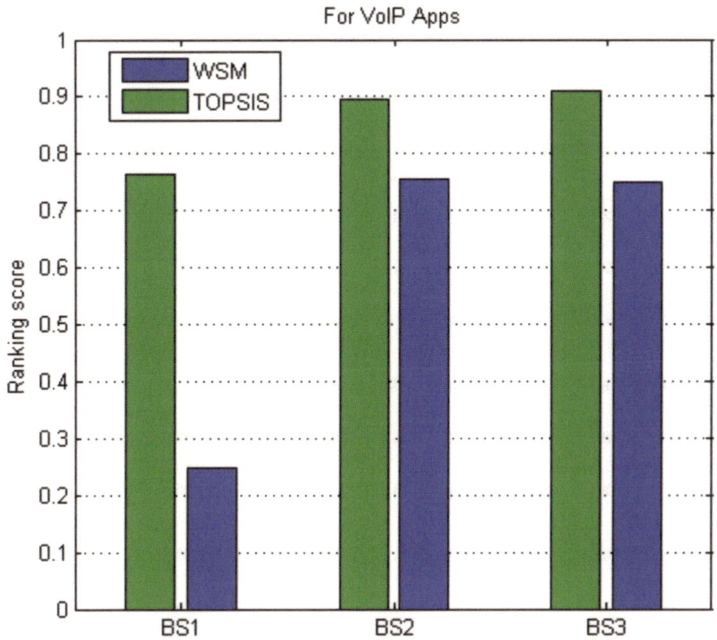

Figure 4. The WSM and TOPSIS scores for real-time applications

From table 4, we can see that application requirement or the weighting factor has an influence on the ranking order. As mentioned before, both WSM and TOPSIS have different ranking orders for VoIP as a real-time applicatipn. This is further exemplified by the sensitivity analysis on the weighting factor in Figure 6 and Figure 7 respectively.

When the congention delay criteria weighting is changed, and other parameters are kept constant we found that the ranking result is more sensitive when TOPSIS method is used. For example, in Figure 6 using WSM method the BS2 score changes form 0.6 to 0.95 when the congestion delay creiteria weighting is changed, so the range of the score change is 0.35. Whereas in Figure 7 using TOPSIS method when the congention delay criteria weighing is changed, the BS2 score changes from 0.18 to 0.92, so the range of the score change is 0.74, and form this it is clear that TOPSIS method is more sensitive to the weighting factor.

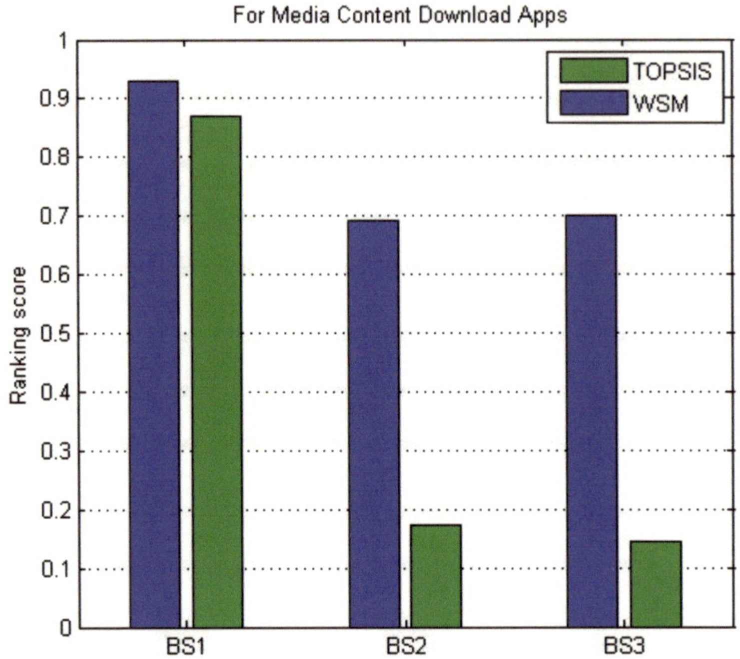

Figure 5. The WSM and TOPSIS scores for the non- real-time application

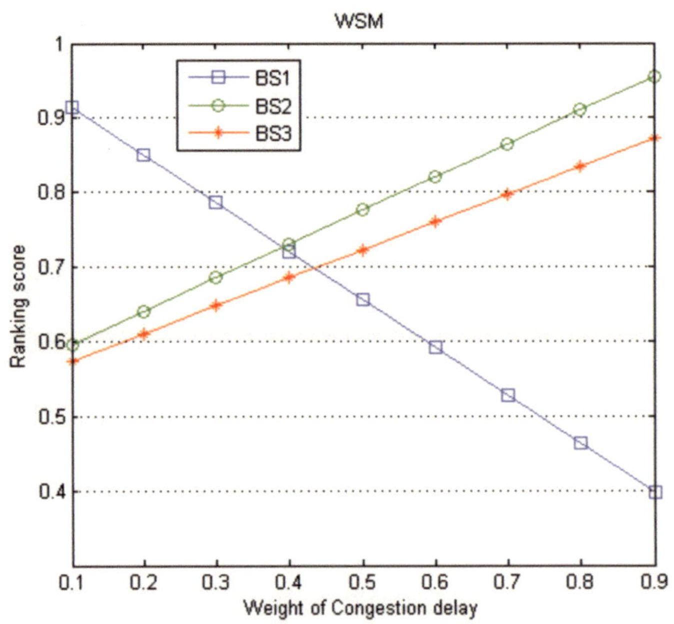

Figure 6. Sensitivity of congestion delay weighting using WSM

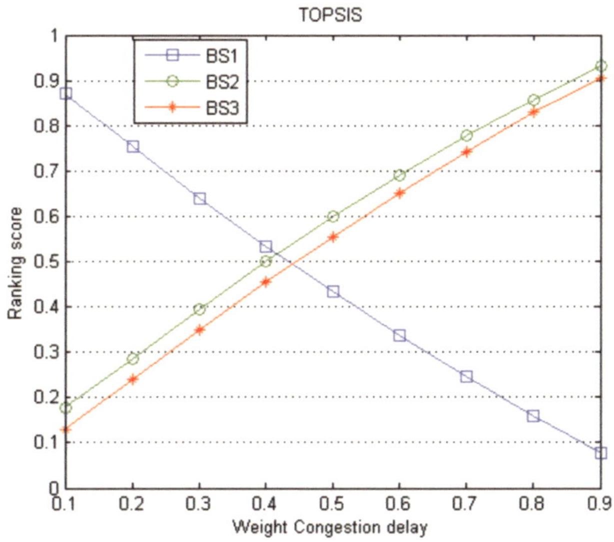

Figure 7. Sensitivity of congestion delay weighting using TOPSIS

6. Conclusion

Mobile WiMAX is one of the 3G/4G broadband wireless technology network that is capable of delivering triple play (voice, data, and video) services. The variety of user application needs different network requirements. So, handover mechanism and the cell selection scheme have to be efficient to meet the application need during and after the handover process. This paper described a smart way of TBS selection. It is based on multi-criteria based selection to meet the user requirements during handover. We proposed the AHP method for criteria weighting and TOPSIS for alternative BSs ranking as a multi-criteria decision-making selection scheme to meet the MS application requirements. The proposed, Hybrid AHP and TOPSIS-based Cell Selection (HATSC) scheme is based on multi-criteria such as CINR, bandwidth (BW) and congestion delay to select the TBS. Numerical analysis shows that TOPSIS is more sensitive to criteria weighting factor, and WSM gives a relative conservative ranking result, and both of them results in better performance than the conventional signal strength scheme.

Author details

Mohammed A. Ben-Mubarak, Borhanuddin Mohd. Ali, Nor Kamariah Noordin, Alyani Ismail and Chee Kyun Ng

Department of Computer and Communication Systems Engineering, Faculty of Engineering, Universiti Putra Malaysia, Selangor, Malaysia

References

[1] Mohamed, M, Zaki, F, & Mosbah, R. Improving Quality of VoIP over WiMAX", IJCSI International Journal of Computer Science Issues, (2012). , 9(3), 85-91.

[2] WiMAX Forum members. Mobile WiMAX- Part I: A Technical Overview and Performance Evaluation", March [(2006).

[3] Ben-mubarak, M, Ali, B. M, Noordin, N. K, Ismail, A, & Ng, C. K. Review of Handover Mechanisms to Support Triple Play in Mobile WiMAX". IETE Tech Rev, (2009). , 258-267.

[4] Ben-mubarak, M, Ali, B. M, Noordin, N. K, Ismail, A, & Ng, C. K. Movement Direction based-Handover Scanning for Mobile WiMAX", the 17Asia-Pacific Conference on Communication (APCC 2011), Sabah-Malaysia, Oct. (2011). , 2-5.

[5] IEEE Standard for Local and metropolitan area networks Part 16: Air Interface for Fixed and Mobile Broadband Wireless Access Systems. (2006).

[6] Choi, S, Hwang, G, Kwon, T, Lim, A, & Cho, D. Fast Handover Scheme for Real-Time Downlink Services in IEEE 802.16e BWA System". Proceeding of IEEE Vehicular Technology Conference, (2005). , 2028-2032.

[7] Chen, L, Cai, X, Sofia, R, Huang, Z, & Cross-layer, A. Fast Handover Scheme for Mobile WiMAX". Proceeding of IEEE Vehicular Technology Conference, Baltimore, MD, (2007). , 1578-1582.

[8] Chen, J, Wang, C, & Lee, J. Pre-Coordination Mechanism for Fast Handover in WiMAX Networks". Proceeding of IEEE (AusWireless2007), Sydney, Australia, (2007). , 15-20.

[9] Jiao, W, Jiang, P, & Ma, Y. Fast Handover Scheme for Real-Time Applications in Mobile WiMAX". Proceeding of IEEE International Conference on Communications (ICC'07), Glasgow, Scotland, (2007). , 6038-6042.

[10] Yeh, J, Chen, J, & Agrawal, P. Fast Intra-Network and Cross-Layer Handover (FINCH) for WiMAX and Mobile Internet. IEEE Transactions on Mobile Computing, (2009). , 558-574.

[11] Gu, S, & Wang, J. An enhanced handover target cell selection algorithm for WiMAX network," 15th Asia-Pacific Conference on Communications (APCC), (2009). , 774-777.

[12] Yang, S, Wu, J, & Huang, H. A vertical Media-Independent Handover decision algorithm across Wi-Fi and WiMAX networks", 5th IFIP International Conference on Wireless and Optical Communications Networks, (2008). , 1-5.

[13] Wu, J, Yang, S, Hwang, B, & Terminal-controlled, A. vertical handover decision scheme in IEEE 802.21-enabled heterogeneous wireless networks", Int. J. Commun. Syst. (2009). , 819-834.

[14] Triantaphyllou, E, Shu, B, Sanchez, N, & Ray, T. Multi-criteria decision making: an operations research approach", Encyclopedia of Electrical and Electronics Engineering. New York: John Wiley & Sons, (1998). , 15, 175-186.

[15] Yap, K. All-IP 4G Mobile Networks and Beyond", WINLAB Seminar, Rutgers University, North Brunswick, (2010).

[16] Yap, K, Katti, S, Parulkar, G, & Mckeown, N. Deciphering a Commercial WiMAX Deployment using COTS Equipments", Stanford University, Technical Report, (2010).

[17] Straccia, U. Multi-criteria decision making in fuzzy Description Logics: A First step", In Proceedings of the 13th International Conference on Knowledge-Based & Intelligent Information & Engineering Systems, Lecture Notes in Artificial Intelligence, Springer, (2009). , 79-87.

[18] Ramanathan, R. A note on the use of the analytical hierarchy process for environmental impact assessment", J Environ Manag (2001). , 27-35.

Chapter 5

EAP-CRA for WiMAX, WLAN and 4G LTE Interoperability

E. Sithirasenan, K. Ramezani, S. Kumar and
V. Muthukkumarasamy

Additional information is available at the end of the chapter

http://dx.doi.org/10.5772/54837

1. Introduction

Today we are moving into a "post-PC" world! Not many people sit in front of custom built PCs to do their businesses any more. Hand held devices such as iPod Touch, iPhone, Galaxy S3, iPad, Galaxy Tab, Airbook, Notepad etc. are bringing in a new paradigm as to how people use and communicate information. These devices can be thought as a theoretical "black-box". They are for people who want to use it without wanting to know how they work. Such devices have third generation user interfaces – multi touch, physics and gestures (MPG). They need updates, but the user is not worried of how and where the files are stored. When a new application is installed, the user sees the icon and starts using it. The user is not interested in, what files were installed or where it was installed – there is no file management. The post-PC approach to dealing with software is that it's discovered on an app store, downloaded with a single touch and deleted with another touch. Updates all come at once from the app store and it all happens behind the scene with minimal user involvement. All this is happening and adopted rapidly because people are able to do a number of things without being restricted to one place. They can download apps, watch movies, listen to news, browse the web etc. while on the move.

However, the mobility of these post-PC devices is restricted to some extent due to the limitations in wireless data connectivity. A wireless device at home should preferably get its data connectivity through the wireless router, while on the move from the 3G or 4G network and while at work from the office wireless network. To achieve this interoperability the wireless devices must be recognized by the various networks as it roams from one network to another. Integration of wireless networks has its own advantages and disadvantages. One type of network that is suitable for a particular application may not be appropriate for another. A security mechanism that is effective in one environment may not be effective in the other. There can be situations where different types of networks coexist in one geographical area. However, due to the inherent nature of the wireless communications, wireless networks encounter numerous security problems compared to its wired counterpart. The most significant of these is the first time association. Whether it is a WLAN [1], WiMAX [2] or a 4G LTE [3], all wireless networks will have this setback. The lack of physical connectivity (anchor-attachment) from the wireless device to the network makes the wireless network more vulnerable and hard to protect against authenticity, confidentiality, integrity and availability threats [4][5]. Hence, to overcome this first time association problem wireless devices adopt a range of different techniques.

The Robust Security Network Association (RSNA) proposed in IEEE 802.11i [6] has emerged as the most popular method to counter the first time association problem. The RSNA technique is widely used in both WLANs and WiMAX. Although IEEE 802.11i security architecture offers sufficient protection to the wireless environment, it is up to the implementer to guarantee that all issues are addressed and the appropriate security measures are implemented for secure operation. A single incorrectly configured station could lead the way for a cowardly attack and expose the entire organizational network [7][8].

Notwithstanding the configuration issues, RSNA is the most preferred first time association method for wireless networks. The use of IEEE 802.1x port based access control [9] makes it more flexible for mutual authentication and key distribution. However, RSNA does not provide options for coordinated authentication in a heterogeneous network environment. This results in the wireless users having to use different credentials to authenticate with different wireless networks. Hence, a wireless device will have to repeatedly authenticate itself as it roams from one network to another operators' network, be it the same type of network or different. Therefore, a Coordinated Robust Authentication (CRA) Mechanism with the ability to use a single set of credentials with any network, wireless or wired would be of immense significance to both network users and administrators. In this chapter we present technical details of CRA together with some experimental results. However, before illustrating the details of CRA, we first present an overview of RSNA.

1.1. Robust security network association

The IEEE 802.11i standard defines two classes of security framework for IEEE 802.11 WLANs: RSN and pre-RSN. A station is called RSN-capable equipment if it is capable of creating RSN associations (RSNA). Otherwise, it is a pre-RSN equipment. The network that only allows RSNA with RSN-capable equipments is called an RSN security framework. The major difference between RSNA and pre-RSNA is the 4-way handshake. If the 4-way handshake is not included in the authentication / association procedures, stations are said to use pre-RSNA. The RSN, in addition to enhancing the security in pre-RSN defines a number of key management procedures for IEEE 802.11 networks. It also enhances the authentication and encryption mechanisms from the pre-RSN. The enhanced features of RSN are as follows:

Authentication Enhancement: IEEE 802.11i utilizes IEEE 802.1X for its authentication and key management services. The IEEE 802.1X incorporates two components namely, (a) *IEEE 802.1X Port* and (b) *Authentication Server (AS)* into the IEEE 802.11 architecture. The IEEE 802.1X port represents the association between two peers as shown in Figure 1. There is a one-to-one mapping between IEEE 802.1X Port and association.

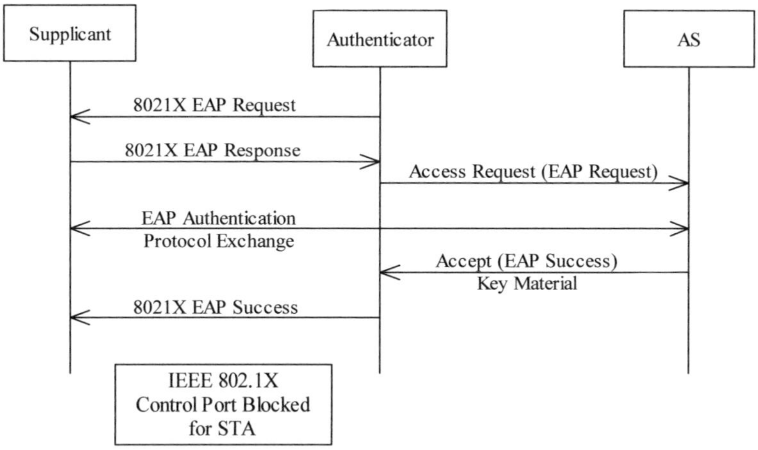

Figure 1. IEEE 802.1X EAP Authentication

Key Management and Establishment: Two ways to support key distribution are introduced in IEEE 802.11i: *manual key management* and *automatic key management.* Manual key management requires the administrator to manually configure the key. The automatic key management is available only in RSNA. It relies on IEEE 802.1X to support key management services. More specifically, the 4-way handshake is used to establish each transient key for packet transmission as in Figure 2.

Encryption Enhancement: In order to enhance confidentiality, two advanced cryptographic algorithms are developed: Counter-Mode/CBC-MAC Protocol (CCMP) and Temporal Key Integrity Protocol (TKIP). In RSN, CCMP is mandatory. TKIP is optional and is recommended only to patch any pre-RSN equipment.

During the initial security association between a station (STA) and an access point (AP), the STA selects an authorized Extended Service Set (ESS) by selecting among APs that advertise an appropriate Service Set ID (SSID). The STA then uses IEEE 802.11 Open System authentication followed by association to the chosen AP. Negotiation of security parameters takes place during association. Next, the AP's Authenticator or the STA's Supplicant initiates IEEE 802.1X authentication. The Extensible Authentication Protocol (EAP) used by IEEE 802.1X will support mutual authentication, as the STA needs assurance that the AP is a legitimate Access Point.

The last step is the key management. The authentication process creates cryptographic keys shared between the IEEE 802.1X AS and the STA. The AS transfers these keys to the AP, and the AP and STA use one key confirmation handshake, called the 4-Way Handshake, to complete security association establishment. The key confirmation handshake indicates when the link has been secured by the keys and is ready to allow normal data traffic.

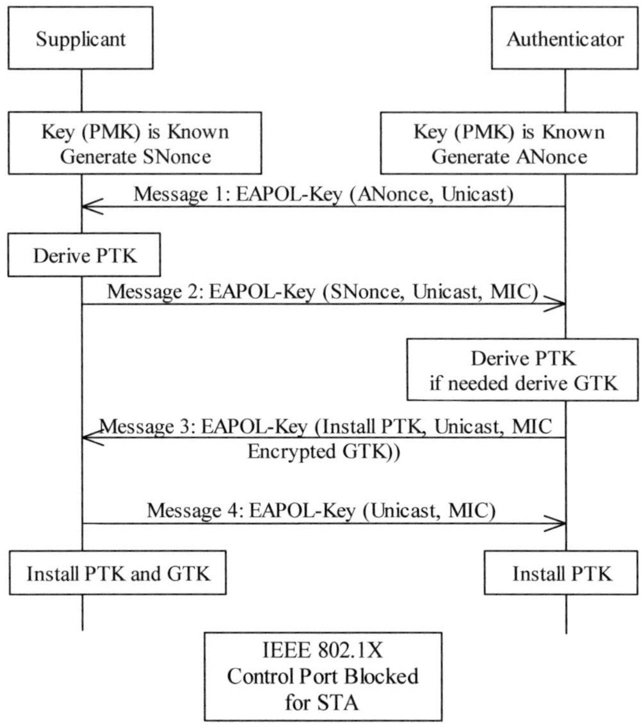

Figure 2. Establishing pairwise & group keys [6]

In the case of roaming, an STA requesting (re)association followed by IEEE 802.1X or pre-shared key authentication, the STA repeats the same actions as for an initial contact association, but its Supplicant also deletes the PTK when it roams from the old AP. The STA's Supplicant also deletes the PTKSA when it disassociates / de-authenticates from all basic service set identifiers in the ESS. An STA already associated with the ESS can request its IEEE 802.1X Supplicant to authenticate with a new AP before associating to that new AP. The normal operation of the DS via the old AP provides communication between the STA and the new AP.

2. Existing methods for integrating wireless networks

Iyer et al. [10] claim that WLAN and WiMAX are particularly interesting in their ability towards mobile data oriented networking. They confirm that a scheme enabling mobility across these two would provide several advantages to end-users, wireless operators as well as Wireless Internet Service Providers (WISP). Further, they propose a technique with a common WLAN/WiMAX mobility service agent for use across WLAN and WiMAX access. By incorporating an acceptable mapping mechanism between WLAN and WiMAX, they interface a WLAN Access Point with the WiMAX Access Service Network (ASN) gateway. The mapping

http://dx.doi.org/10.5772/54837

function inside WLAN access point maps all 802.11 events to the WiMAX events. For example the event association request will be mapped to WIMAX pre-attachment request.

In their architecture the problem of handling mobility across WLAN and WiMAX boils down to the problem of handling mobility across WiMAX base stations that already have concrete solutions. Also, the mapping function consumes 1.82 seconds for EAP-TLS authentication in comparison to few milliseconds in CRA. Further, their proposed architecture enables the same IP address to be used across both the WLAN and the WiMAX network interfaces, and keeps it seamless from an application perspective.

Distributed authentication scheme proposed by Machiraju et al. [11] relies on Base Stations (BS) to collectively store authentication information. To achieve the goal of single point of access they introduce the notion of tokens. The token contains the identity and other information regarding the user. Each mobile user has exactly one token that is stored at the base station where the mobile user is receiving service. When the mobile user moves between base stations, its token moves along with the user, thus, eliminating the need to maintain costly infrastructure required by traditional centralized scheme. They assert two main disadvantages of centralized authentication methods. Firstly, a server must be available. Without a server the authentication process cannot be completed. Secondly, there must be a highly reliable backhaul. The latter is due to the authentication process creating a large volume of traffic, usually of a higher priority than normal traffic. They further emphasize that their scheme is optimized for mobility-induced handover re-authentication and, thereby reducing the authentication overheads. This study however, does not clarify how the base stations will initiate contact with each other. The security approach to establish a secure connection between the BS is not determined. Moreover the details to establish trust between base stations and actions taken in case of base stations being compromised are not provided. The capabilities required to perform the expected functionality of a BS are not addressed.

The EAP-FAMOS authentication method developed by Almus et at. [12] use the Kerberos based authentication in the existing EAP framework. It allows secure and true session mobility and requires the use of another EAP method, only for the initial authentication. It uses the keying material delivered by the other EAP method during the initial authentication for its Kerberos-based solution for fast re-authentication. Mobility is based on Mobile IPv4 and a sophisticated handover supported by a so-called Residential Gateway together with a Mobility Broker located in the ISP's backend network. Their performance studies show that Wi-Fi technology can be used in mobile scenarios where moving objects are limited to speeds below 15kmh. Further, they state that applications requiring very low delay and allowing only very short service interruptions can be supported by their technique.

OSNP is another EAP method based on Kerberos proposed by Huang et al. [13]. The protocol provides intra-domain and inter-domain authentication to a peer that already has its security association with the home network. The authors have proposed a hierarchal design for KDC servers with the Root KDC responsible for providing directory service to other KDC servers. In case of a request to a particular network other than the peer's Home network, the authentication server in the new network will obtain the authenticity of the peer from the home KDC. Although the authors suggest a quick password based authentication and roaming mechanism, they fail to provide details of the hierarchical design of KDC servers and the agreement between them. Moreover, all servers share a group key and in case of a key compromise, access points can masquerade as legitimate authenticators.

Apart from the high administrative costs in Kerberos based methods; their solution is mainly targeted at specific wireless networks and authentication mechanisms. Wireless service providers use different authentication schemes on their diverse types of wireless networks. For example, a WiMAX service provider may use the EAP-TLS authentication scheme on their custom Authentication Authorization and Accounting (AAA) server, whereas corporate entities may want to use EAP-TTLS authentication mechanism facilitating the use of their existing authentication databases such as Active Directory, LDAP, and SQL. Hence, for convergence of wireless networks it is significant to develop an authentication mechanism that is versatile and simple so that it can be effectively used in any type of wireless network.

Narayanan et al. [14] propose ERP, an extension to the EAP framework and an EAP key hierarchy to support Re-authentication. As specified in RSNA, MSK is generated on successful completion of the authentication phase (phase 2 of RSNA). Subsequently MSK is passed to the authenticator to generate the TSK (phase 3 of RSNA). The TSK is then used for data encryption between the supplicant and the authenticator. However, the EAP framework proposed by Narayanan et al. suggests two additional keys to be derived by all EAP methods: the Master Session Key (MSK) and the Extended MSK (EMSK) which forms the EAP key hierarchy. They make use of the EMSK for re-authentication and successive key derivations.

ERP defines two new EAP messages EAP-Initiate and EAP-Finish to facilitate Re-authentication in two round trip messages. At the time of the initial EAP exchange, the peer and the server derive an EMSK along with the MSK. EMSK is used to derive a re-authentication Root Key (rRK). The rRK can also be derived from Domain-Specific Root Key (DSRK), which itself is derived from the EMSK. Further, a re-authentication Integrity Key (rIK) is derived from the rRK; the supplicant and the authentication server use the rIK to provide proof of possession while performing an ERP exchange. After verifying proof of possession and successful authentication, re-authentication MSK (rMSK) from the rRK is derived. rMSk is treated similar to MSK obtained during normal EAP authentication i.e. to generate TSK [15].

Apart from the few modifications to the EAP protocol due to the introduction of two new EAP codes, ERP integrates with the existing EAP framework very well. To demonstrate the possession, supplicant uses rIK to compute the integrity checksum over the EAP-Initiate message. The algorithm used to compute integrity checksum is selected by the peer and in case of server's policy does not allow the use of cipher suite selected by the peer; the server sends a list of acceptable cipher suites in the EAP-Finish / Re-auth message. In this case the peer has to re-start the ERP process by sending the EAP-Initiate message and the integrity checksum using the acceptable cipher suites. Furthermore ERP also recommends use of IPsec or TLS to protect the keying materials in transit. However, EAP-ERP requires a full EAP authentication at first when a user enters a foreign network. Further, if one supplicant for any reason has not been able to extract domain name of the foreign network then it should solicit it from its Home server, this can result in long authentication delays.

http://dx.doi.org/10.5772/54837

Increasing use of Mobile devices and new data capabilities on these devices suggest more attention for fast and secure handover. Authentication mechanisms such as EAP-AKA and EAP-SIM facilitate handover and re-authentication for 3GPP interworking.

3. Coordinated Robust Authentication

The principal notion behind the Coordinated Robust Authentication (CRA) [16] mechanism is that every wireless device will primarily be associated with one wireless network, which can be referred to as its HOME network. The credentials used by a wireless device to associate with its HOME network are assumed to be robust and specific to that network. Therefore, a wireless device must be able to use its authority in the HOME network to reliably associate with any other FOREIGN network. In this context, the AAA server that authorizes the wireless device in its home network is called as the HOME AAA Server and the AAA server in a foreign network is called as the FOREIGN AAA Server. Hence, in CRA, a wireless device will require only one set of credentials that it uses to access the home network to access any type of foreign networks. CRA considers both different types of networks and different authentication mechanisms that may be specific and effective to that type of network.

Therefore, in this mechanism a wireless device will deal with one HOME network and a number of FOREIGN networks. It also assumes that the security mechanism used in the HOME network is the most effective that can be adapted to the type of wireless devices used in the network. Further, it is assumed that the HOME AAA server will have pre-arranged agreements with the FOREIGN AAA servers for secure communications by other means such as IPSec, SSL etc.

Figure 3 outlines the messages exchanged in CRA. As in the RSNA, the CRA also includes a discovery phase that comprises of the six 802.11 open system association messages. During this phase a wireless device that is in the FOREIGN network will advertise that it is capable of EAP-CRA together with other allowed EAP methods. Hence, an authenticator in the FOREIGN network can initiate EAP-CRA if it is capable of managing it. Once they both agree on the EAP-CRA mechanism, the authenticator can initiate the EAP-CRA by sending the EAP Request / Identity message to the supplicant (message 7 in Figure 3). The supplicant in return will reply with the EAP Response / Identity message (message 8). The Response / Identity message is passed to the FOREIGN AAA server as a RADIUS Access Request message. At this stage unlike in the other EAP authentication methods the AAA server will pass the Access Request message to the relevant HOME AAA server for validation. If the HOME AAA server successfully validates the Identity information sent by the wireless device, it then responds with an Access Accept message with the necessary keying material to the FOREIGN AAA server. The keying material, in-turn, is passed to the authenticator with the RADIUS Access Accept message. The authenticator can then use the keying material to initiate the 4-way handshake process to generate the TSK. Further details of the CRA protocol are explained in the next section.

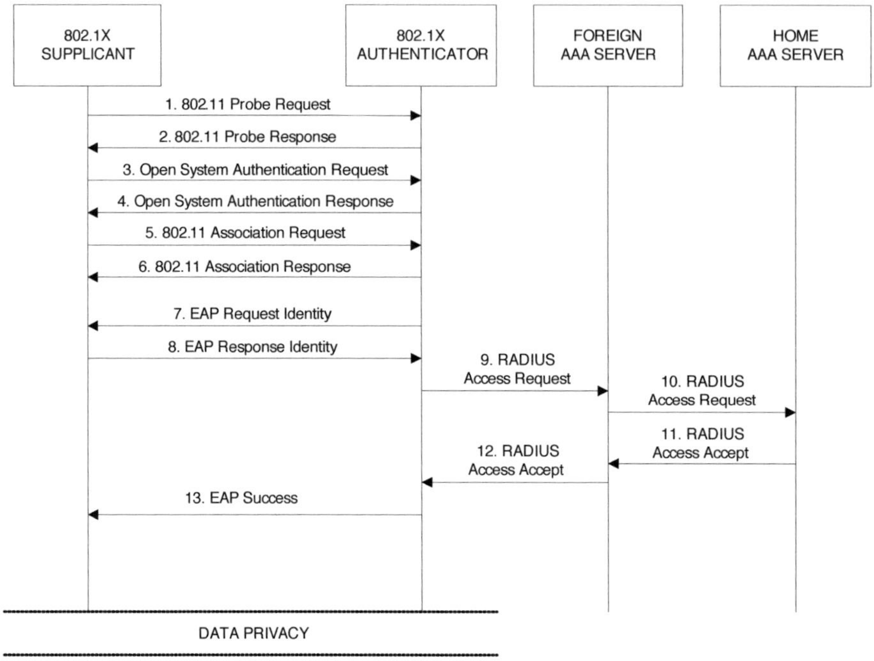

Figure 3. Coordinated authentication message exchange

3.1. The EAP-CRA protocol

With regard to mutual authentication EAP-CRA uses RADIUS servers as suggested in IEEE 802.1x [17]. RADIUS protocol exhibits better performance compared to other mutual authentication protocols [18]. EAP-CRA offers direct communication between radius servers by pre-arranged agreement or the servers could find each other dynamically. In case the RADIUS servers do not have a pre-arranged agreement then they can use their CA-signed PKI certificates to ascertain trust between servers.

All AAA servers that participate in the EAP-CRA must have some pre-arranged agreement for secure communication. Assuming that all AAA Servers that participate in the EAP-CRA are in possession of their CA-signed PKI certificates, the CRA protocol uses the CA-signed PKI certificates to communicate between the FOREIGN and the HOME AAA servers. However, other options for secure communications such as a virtual private network (VPN) or SSL can also be used. In the protocol details shown in Figure 4, CRA uses the already available CA-signed PKI certificates of the FOREIGN and the HOME AAA servers for secure communication. Message 3 is encrypted using the private key of the FOREIGN AAA server ($E_{KP_F}[HostName,\ E_{KU_H}[EMSKname,\ SeqNo.]]$) and message 4 is encrypted using the public key of the FOREIGN AAA server ($E_{KU_F}[DSRK]$). However, in Figure 4, we have left the issue of secure communication between the FOREIGN and the HOME AAA server open, to confirm that other options are possible.

According to the EAP-CRA protocol, in response to the EAP-CRA Request Identity message (message 1 in Figure 4), the supplicant sends an EAP Response message with its *Identity* (EMSKname and Sequence number) encrypted with the public key of the HOME AAA server (message 2 in Figure 4) along with the unencrypted host name of the HOME AAA server. EMSKname is used to identify the corresponding EMSK and Sequence Number for Replay protection by the Home AAA server. The authenticator, having received the encrypted *Identity* will pass it to the FOREIGN AAA server as it is. The FOREIGN AAA server uses the fully qualified *Host Name* provided in EAP-CRA Response message to determine the Home AAA server. The FOREIGN AAA server will append its *Domain name* to the received message (EAP-CRA Response) and pass it to the HOME AAA server using the secure method described above (message 3).

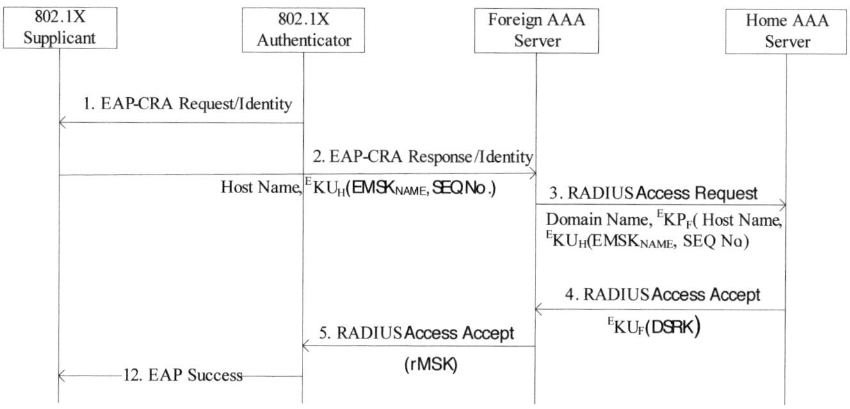

Figure 4. Coordinated Robust Authentication (CRA) Protocol.

The HOME AAA server will then have to do a double decryption to find the identity of the HOME wireless device. If the wireless device is positively identified, the HOME AAA server calculates *DSRK* (Domain Specific Re-authentication key). DRSK is calculated using *Domain Name* as an optional data in the key derivation specified in [15]. HOME AAA server will then send the *DSRK* to the FOREIGN AAA server after encrypting the message using the public key of the FOREIGN AAA server (message 4). This process is illustrated in Figure 5. The FOREIGN AAA server can use its private key to decrypt the received message to discover the *DSRK* and generate *rMSK* (Re-authentication Master Session Key). rMSK is calculated using a sequence number as an optional data specified in [14]. The *rMSK* can then be transferred to the authenticator with the RADIUS Access Accept message (message 5 in Figure 4). Finally the authenticator sends the EAP success message to the wireless device indicating the completion of the CRA authentication and the beginning of the key distribution phase.

Two sequence numbers, one with HOME AAA server and one with FOREIGN AAA server is maintained for replay protection of EAP-CRA messages. The sequence number maintained by the supplicant and HOME AAA server is initialized to zero on the generation of EMSK. The server sets the expected sequence number to the received sequence number plus one on every successful Re-authentication request i.e. on generation of DSRK. Similarly the supplicant and the FOREIGN AAA server maintain a sequence number with the generation of rMSK until the supplicant is in the FOREIGN AAA server's domain.

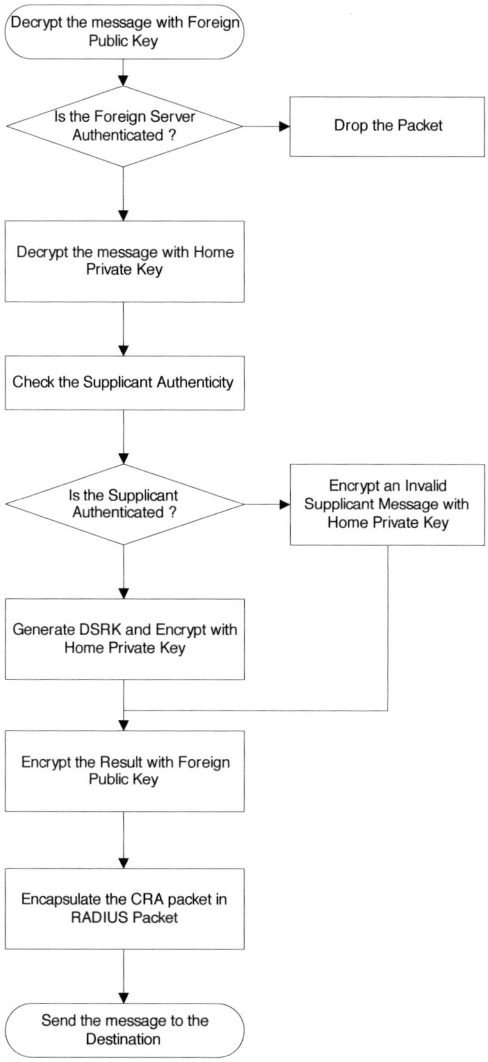

Figure 5. EAP-CRA on Home Server

On receiving the EAP success message, the peer generates rMSK independently leading to the key distribution phase. The key distribution phase will be similar to that of the RSNA where the supplicant and the authenticator will use the MSK to derive TSK. Once the Temporary Session keys (TSK) are derived normal data communication can commence. In the next section we discuss the server side communication of the CRA authentication mechanism.

3.2. Extentions to RADIUS

EAP-CRA uses RADIUS as the transportation protocol between the Home and Foreign servers. However the RADIUS protocol is a client-server protocol. The RADIUS server, when forwarding the authentication packet to another RADIUS server, designates the sender as client. Hence, the foreign server's only responsibility is to fulfill the role of a proxy server and to forward the RADIUS packets to the Home server. EAP-CRA takes advantage of RADIUS communication and encapsulates the EAP-CRA messages inside the RADIUS packets. There are two viable approaches to designing the security methods that were discussed in the previous section.

The first approach is to implement the security features inside the attribute field of the RADIUS packet (Table 1). The attribute field of each RADIUS Packet includes at least three fields that enable the RADIUS packet to carry EAP messages or other information for Dial in user. The attribute field can be used to encapsulate EAP-CRA messages inside the RADIUS packet. Extensions to RADIUS protocol so far proposed have been for the purpose of modifying or creating new attributes such as EAP or apple extensions for RADIUS, each of which has particular attributes.

0 0	1	2	3	4	5	6	7	8	9	1 0	1	2	3	4	5	6	7	8	9	2 0	1	2	3
Type								Length								Value ...							

Table 1. Attributes in a RADIUS packet

Type 79 is for EAP messages and 92-191 are Unused. If the value is string or text type then the length can be from 1 to 253 octets. Therefore the type value can be between 92 to 191 octets for the EAP method. The type of the value will be string and as with other EAP methods data is encapsulated inside the RADIUS packet. The foreign server can encapsulate the encrypted message inside the RADIUS packet, so that the home server must first decrypt the message and then respond by a proper RADIUS message to the foreign server.

The second approach is to use a dependent VPN over a SSL connection between the two servers prior to RADIUS communication. The RADIUS packets can then be sent in a secure channel. However, EAP-CRA does not use this method because it entails extra network administration. It also creates a connection delay prior to the EAP-CRA message transmission. Also, the use of PKI actually provides a more secure channel by which the EAP-CRA message can be sent and received.

3.2.1. EAP-CRA message and process details

The proposed EAP-CRA packet is depicted in Table 2. The reasons for designing each of the fields are illustrated based on the associated requirements. The fields are transmitted from left to right. The first influencing factor of EAP-CRA is that it is based on the EAP protocol.

Therefore, the fields, code, identifier and length are inherited from an EAP structure. The explanation of each field is listed below.

<table>
<tr><td>0
0</td><td>1</td><td>2</td><td>3</td><td>4</td><td>5</td><td>6</td><td>7</td><td>8</td><td>9</td><td>1
0</td><td>1</td><td>2</td><td>3</td><td>4</td><td>5</td><td>6</td><td>7</td><td>8</td><td>9</td><td>2
0</td><td>1</td><td>2</td><td>3</td><td>4</td><td>5</td><td>6</td><td>7</td><td>8</td><td>9</td><td>3
0</td><td>1</td></tr>
<tr><td colspan="8">Code</td><td colspan="8">Identifier</td><td colspan="16">Length of CRA</td></tr>
<tr><td colspan="8">Type</td><td colspan="8">Flags</td><td colspan="16">CRA Message Length</td></tr>
<tr><td colspan="16">CRA Message Length</td><td colspan="16">CRA Data ...</td></tr>
</table>

Table 2. CRA Packet

The Code field is one octet and identifies the type of EAP packet. EAP Codes are assigned as 1 for Request, 2 for Response, 3 for Success and 4 for Failure. The Identifier field is one octet and aids in matching responses with requests. The Length field is two octets and indicates the length of the EAP packet including the Code, Identifier, Length and Data fields. Octets outside the range of the Length field should be treated as Data Link Layer padding and should be ignored on reception. The Flags field includes the following fields:

0	1	2	3	4	5	6	7
L	M	S	T	R	R	R	R

L = Length included, M = More fragments, S = EAP-CRA start, R = Reserved, T = Source Type

Table 3. Add Caption

3.2.2. Two kinds of RADIUS packets in EAP-CRA

In EAP-CRA, RADIUS packets are divided into two categories, based on their content. The first category includes those messages sent from an access point to the foreign server and the second type is those exchanged between a Home and Foreign server. In the first scenario, the supplicant encrypts the EAP-CRA message using the Home server public key and sends it to the foreign server. Between the home server and the client, the authenticator encapsulates the message inside a RADIUS packet and sends it to the foreign server. On the other hand, when the two servers are in communication with each other they sign the EAP-CRA message first using their own private key and then by encrypting the message using the other server's public key. Therefore, the content of the RADIUS packets differ depending on whether they are received from an authenticator or from an authentication server. The field T in the fragmentation field is for source type of the packet. If the packet is from or is sent to an authenticator then the value will be set to 0. Otherwise, if the source is a server, then the value will be set to 1.

Retry behavior: It is possible during peer communication that a response will not occur within the expected time. In which case, there must be a way to specify how many messages will be sent to make sure that another peer is not present. The time to resend the message is another parameter which needs to be determined. The exact number for the time and trials will be decided in the actual implementation and depends on the protocol process time, line traffic and other unforeseen factors. One of the issues present in retry is the duplicate packets which must be handled by the receiving peer. Three retries will be performed, forming the base configuration for the EAP-CRA.

Fragmentation: EAP-CRA message may span multiple EAP-packets due to the multiple public and private key encryptions; hence there must be a method, to be engineered in the servers, for handling the fragmentation. As a base for work on the fragmentation, the length of the TLS record can be up to 16384 octets, while the TLS message may be 16 MB if it carries the PKI certificate of a server. However, to protect against denial of service attacks and reassembly lockup there must be maximum size set for the group of the fragmented messages. An example can be seen in what was implemented for EAP-TLS[19]. The exact numbers will be determined during implementation of the protocol, and will reveal the average length of long EAP-CRA messages. For the purposes of initial configuration, this number can be borrowed from EAP-TLS which is 64 KB.

Since EAP is an uncomplicated ACK-NAK protocol, fragmentation support can be provided according to a relatively simple process. Damage or loss of fragments during transit is an inevitable risk for any communication. In EAP, these fragments will be retransmitted, and because sequencing information is included in EAP's identifier field, a fragment offset field like that of IPv4 is not necessary.

EAP-CRA fragmentation support will be provided by adding flag fields to the EAP-CRA packets inside the EAP-Response and EAP-Request. Flags include the Length (L), More fragments (M), and Start (S) bits. The L flag indicates the presence of the four octet Message Length field. It *must* be set in the first piece of a fragmented EAP-CRA message or set of messages. The M flag will be set in all except the last fragment showing that there are more frames to follow. The S flag will only be for the EAP-CRA start message sent from the EAP server to the peer. The T flag refers to the source type of the EAP-CRA message; whether it is coming from an 802.1x authenticator or from an authentication server. If there is a fragmented message, both server and the other peer must acknowledge the receipt of a packet with the flag set to M. The response can be an empty message to the other peer showing that the message has not been received.

3.3. Experiments

For our experiments we setup three different scenarios to compare the time taken to authenticate a user. Edu-roaming, EAP-CRA and direct authentication with a single RADIUS authentication server were considered. RADIUS servers were installed on Windows 2003 Server standard edition and all platforms had 2 GB RAM and 2GHz dual core CPU.

Microsoft Internet Authentication Service (IAS) with Microsoft EAP-PEAP was used in these experiments. IAS is the Microsoft implementation of a Remote Authentication Dial-In User Service (RADIUS) server and proxy in Windows Server 2003. As a RADIUS server, IAS performs centralized connection authentication, authorization and accounting for many types of network access including wireless and VPN connections. As a proxy, the IAS forwards authentication and accounting messages to other RADIUS servers.

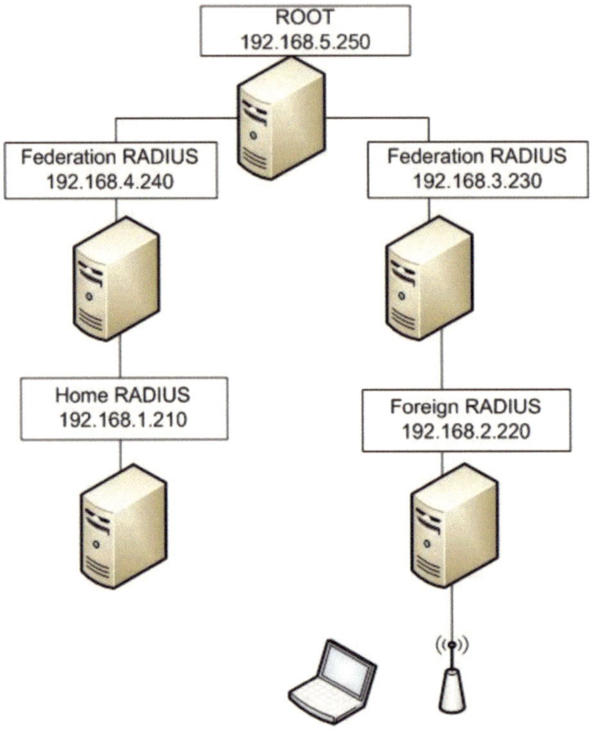

Figure 6. Experimental Edu-roam Setup on LAN

To start with fair baselines both EAP-CRA and Edu-roaming were implemented in LAN but in different IP subnets. Moreover to magnify the delay of authentication for Edu-roaming another setup on Internet was also implemented. The first topology is the Edu-roaming model. Since this is a proprietary model it was implemented on five Microsoft IAS that was installed on the Java virtual box. Because the Edu-roaming has federation level RADIUS servers and one root RADIUS server, we implemented five RADIUS servers in all. Two of the RADIUS servers were for the home and the foreign networks, two as the federation level RADIUS servers and the last one as the Root authentication server. Figure 6 shows the topology for Edu-roaming that was implemented by us.

The second scenario was an implementation of Edu-roaming and EAP-CRA servers on the Internet. Five servers were installed at various remote sites in Brisbane Australia. In all scenarios, the time difference between the first RADIUS request message and the last RADIUS accept message was used for comparing the time taken for authentication. Tables 4 and 5 lists the average times obtained on the LAN and Internet implementations over forty different trials.

Topology	Edu-roam	EAP-CRA	Direct
Average Time (ms)	259	148	119

Table 4. Average Authentication Time on LAN

Topology	Edu-roam	EAP-CRA
Average Time (ms)	4176	750

Table 5. Average Authentication Time on Internet

According to Table 1 there is a 111 milliseconds time difference in the authentication times between Edu-roaming and the EAP-CRA. As explained earlier the EAP-CRA directly communicates with the foreign RADIUS server. Moreover, the difference in authentication times between the CRA approach and direct authentication with the RADIUS server is 29 milliseconds. Table 5 shows the authentication times over the Internet. Here, the RADIUS servers are located at different locations and are connected over the Internet. In this case there is a significant difference in authentication times between Edu-roaming and EAP-CRA approaches. The Edu-roaming approach is almost three times slower than the EAP-CRA approach in this case.

3.4. Discussion

Figure 7 confirms the potential of the EAP-CRA approach compared to the other methods. The main advantage of the EAP-CRA authentication mechanism is the use of only two messages to authenticate a wireless device in a FOREIGN network. Although the time taken between the FOREIGN AAA server and the HOME AAA server may vary depending on the traffic and/ or capacity of the wired network, the use of only two messages in a FOREIGN network makes CRA authentication mechanism very much reliable compared to other available techniques. Further, even if the foreign network uses a less secure authentication mechanism, it still will not affect the EAP-CRA supplicants since their PMKs are supplied by the HOME AAA servers not-withstanding the limitations of the foreign network.

Another significant advantage of the EAP-CRA is its reliance on the HOME security credentials to secure its clients in the foreign network. Hence, it can be assured that the EAP-CRA clients will have the same security guarantee as in their home network in the foreign network. Further, in the case of EAP-TLS authentication with CA-signed PKI certificates, clients will need only a single set of certificates signed by the CA accepted by the HOME AAA server. There will be no need for clients to carry a number of different certificates to authenticate with different networks. Hence, in this context, the EAP-CRA facilitates EAP-TLS authentication and makes it more practical and viable.

Although there are many other techniques proposed for distributed authentication, the advantages of the EAP-CRA technique is its simplicity, robustness and versatility. Unlike many other systems that require additional components such as a token management system or federation of RADIUS servers, the EAP-CRA system depends only on the existing infrastructure, hence, assuring simplicity. The use of existing CA-signed PKI certificates without necessitating other authentication mechanisms such as tokens or smart cards enables the EAP-CRA system to be confined. Further, EAP-CRA system is not limited to WLAN or WiMAX, it can be effectively used with any wireless network, harnessing the unique security features of that particular wireless network. Furthermore, the authentication mechanism (EAP-TLS, EAP-TTLS, EAP-PEAP etc.) used by the wireless network does not influence the EAP-CRA system because it does use any form of mappings between these protocols and the EAP-CRA protocol.

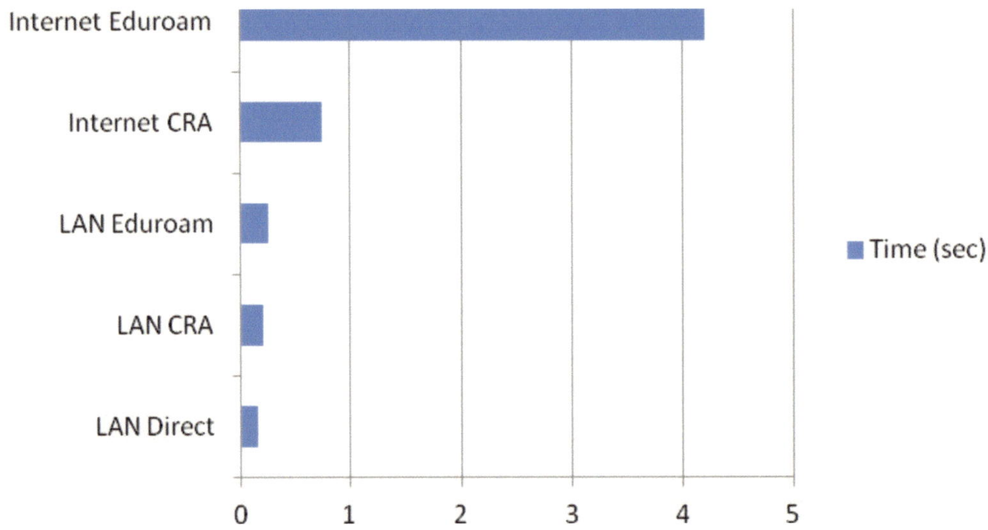

Figure 7. Comparison of Authentication Times

The above discussions illustrate the significance of the CRA approach and emphasize the need for a fast authentication mechanism as opposed to a hierarchical mechanism like the Edu-roam. Although Microsoft IAS provides a similar infrastructure to that of EAP-CRA, it is restricted to Microsoft EAP-PEAP authentications. In contrast EAP-CRA does not rely on any particular authentication protocol. It is designed to reap the maximum leverage of the authentication mechanism that is best for the particular home environment. Hence, when a hand-held device roams in a foreign network it will have the same security guarantee as in the home network.

EAP-CRA is differentiated by other EAP methods in the aspects of communication scope by covering both the foreign and the home authentication servers. Other EAP methods such as EAP-TLS or EAP-TTLS do not consider server to server communication. EAP-CRA provides authentication and communication privacy between the foreign and the home authentication servers based on public key infrastructure. The home and foreign servers have got the public certificates of each other. EAP-CRA encrypts the authentication message twice and then sends it to the other foreign server ensuring privacy and authenticity of the message. Any message from home server will first be signed by the home server's private key and then by the foreign servers public key. Same process happens if the foreign server sends a message to the home server. The signature of a server by the private key authenticates the server to the other server and the public key encryption ensures privacy of the transmitted message. To implement the transmitting of the messages between two authentication servers EAP-CRA suggests using of RADIUS protocol by creating a new attribute field which encapsulates the EAP-CRA message. The EAP-CRA message is the double encrypted message which will be located in the value filed of the RADIUS attribute.

On the negative aspect, the effectiveness of EAP-ERP will depend on the mutual trust established between the participating AAA servers. If the AAA servers do not have any form of prior agreement, it will be up to the discretion of a FOREIGN AAA server whether to accept or deny an EAP-CRA request.

4. Enhancements to EAP-CRA

The Enhanced CRA protocol provides authentication in two modes; Full Authentication and Re-Authentication. With regard to mutual authentication CRA uses RADIUS servers as suggested in IEEE 802.1x. CRA suggests direct communication between radius servers by pre-arranged agreement or the servers could find each other dynamically. In case the RADIUS servers do not have a pre-arranged agreement they can use their CA-signed PKI certificates to ascertain trust between servers.

All AAA servers that participate in the CRA must possess a CA-signed PKI certificate and be capable of obtaining the CA-signed PKI certificates of other participating AAA servers. Assuming that all AAA Servers that participate in the CRA are in possession of their CA-signed PKI certificates, the CRA protocol can communicate between the FOREIGN and the HOME AAA servers securely.

4.1. Full EAP-CRA authentication

Initial assumption of the CRA protocol is that each mobile Node is primarily associated with a Network, which in this context is referred to as the Home network. The security of the Home network and the authentication mechanism used must be robust. It is assumed that an EAP method such as EAP-TLS, EAP-PEAP or EAP-TTLS is used in the Home network. Therefore the values for MSKName, MSK, EMSK and the Time To Live (TTL) for these keys are available for the Peer. Since some of the EAP methods utilize CA-signed PKI certificates to authenticate and secure the communication CRA extends it to add more flexibility to certificate based authentication. We have chosen WLAN as the medium to illustrate the components and messaging of EAP-CRA. Firstly, both the peer and the Foreign Access Point (FAP) discover their capabilities and decide on a suitable protocol to authenticate each other. If both parties are capable of EAP-CRA then the FAP will compose an EAP request message to solicit the identity of the Peer. It should be mentioned that the key for hashing function is generated from the EMSK.

In an unknown network, the peer will first check if the TTL of MSK is still valid. Expired MSK will lead to a failed authentication and will prompt a full authentication. The peer will be responsible to do a full authentication with its Home Network to obtain a fresh MSK. On the other hand, if the MSK is valid, the peer generates a random sequence number and encrypts the EMSKname of home network and the sequence number with the public Key of its HAS. The composed EAP-Response message will be sent to the FAP, which contains the encrypted message, Message Authentication Code, the realm of the home network and the random identity of the peer (message b in List 1).

List 1: Messages Exchanged During CRA Full Authentication

a. $FAP \rightarrow MN : EAP_{req}[ID]$Inline Formula

b. $MN \rightarrow FAP : EAP_{res}[Hostname,\ Realm_h,\ \{EMSKname,\ Seq\#\}UK_h,\ MAC]$ Inline Formula

c. $FAP \rightarrow FAS:\ ACC_{req}[Hostname,\ Realm_h,\ \{EMSKname,\ Seq\#\}UK_h,\ MAC]$ Inline Formula

d. $FAS \rightarrow HAS:\ ACC_{req}[Realm_f,\ \{Hostname\}PK_f,\ \{EMSKname,\ Seq\#\}UK_h]$ Inline Formula

e. $HAS \rightarrow FAS:\ ACC_{res}[\{Hostname\}PK_h,\ \{MSK_{CRA},\ EMSK_{CRA}\}UK_f,\ EAPsuccess,\ Seq\#]$ Inline Formula

f. $FAS \rightarrow FAP:\ ACC_{res}[MSK_{CRA},\ Realm_f,\ ReID,\ Seq\#,\ MAC]$ Inline Formula

g. $FAP \rightarrow MN:\ EAP_{req}[Realm_f,\ ReID,\ Seq\#,\ MAC]$ Inline Formula

h. $MN \rightarrow FAS : EAP_{res}[ACK,\ Seq\#,\ \mathrm{MAC}]$ Inline Formula

i. $FAS \rightarrow MN : EAP_{suc}$ Inline Formula

FAP will encapsulate this EAP-Response message inside a RADIUS Packet and forward it to the foreign authentication server. The FAS will also utilize RADIUS for server-to-server communication. However before sending the received message, the FAP will add its domain name and encrypt the MSKname with its Private Key (message d in List 1). This enables the HAS to authenticate the FAS. Upon receiving the message from a foreign network, HAS is able to check if the FAS is authorized based on the domain name of the FAS. The HAS can authenticate the FAS by verifying the contents of the signed message. Peer authentication will be managed by matching the MSKname with MSK, EMSK, Validation of key timer and the number of re-authentication of the peer. If the MSK is valid the HAS can combine the foreign domain name, sequence number and the previous EMSK to generate new CRA-MSK and CRA-EMSK.

After updating the timer and counter values of the MSKname the HAS creates a RADIUS message which holds Access Accept, encrypted values of CRA-MSK and CRA-EMSK with FAS's Public Key, MAC and privately signed message of domain name – MSKname (message e in List 1).

FAS first checks the signed MSKname to validate the HAS, then stores the MSKname and CRA keys. In addition to these it calculates a new timer, counter and random re-authentication ID for local re-authentication in case the peer stays for longer time in the foreign network. These values are CRA_timer, CRA_counter, and CRA_RND. The value of the CRA_timer must be less than the validity time of the initial MSK. Next, the FAS sends CRA_counter, re_id, EMSKname signed with HAS's private key, Foreign realm and CRA-MSK inside a RADIUS packet to FAP (message f in List 1). The CRA-MSK will be utilized for future communication to provide privacy. The rest of the message is sent to the peer (message g in List 1). The peer will be able to authenticate its home server by verifying the signature and can generate CRA-MSK and CRA-EMSK. It then creates a EAP-Response as an acknowledgment with MSKname. The FAS can then compose a EAP-Success message and send it back to the peer.

On receiving the EAP success message, the peer generates rMSK independently leading to the key distribution phase. The key distribution phase will be similar to that of the RSNA where the supplicant and the authenticator will use the MSK to derive the Temporal Session Key (TSK). Once the TSKs are derived normal data communication can commence.

4.2. EAP-CRA re-authentication

In the previous section we described a roaming-enabled authentication mechanism for users who wish to get connected to a new network, using the security credentials that they use in their home network. Although we anticipate relatively faster CRA authentication, in situations where the user continues to work on a foreign network the need for re-authentication is anticipated.

This section will explain the re-authentication process that can occur due to handover within the same network, i.e. when a user moves from one access point to another. The Enhanced CRA full authentication generates CRA-MSK and CRA-EMSK for a secure communication. Possession of these keys by the supplicant and the FAS can quicken the process of re-authentication. The FAS, after the successful authentication of a supplicant distributes the re-authentication identity and the CRA_Counter to the peer. The counter determines the number of re-authentications which can be acceptable.

The process of re-authentication will be initiated by the authenticator with EAP-Request for supplicant ID. In response the supplicant will check the time since last logon to verify the validity of CRA-MSK. In case the key is expired then a valid peer will fall back to request a full EAP-CRA authentication. On the other hand the supplicant sends its re-authentication ID and realm inside Kname-NAI, a random sequence number with a hashed value of the message. The key for the hash can be generated from the CRA-EMSK and sequence number. Here, the need for the sequence number arises to provide immunity against replay attacks. The authenticator will then forward the EAP-Response encapsulated as a RADIUS packet to the FAS (message c in List 2).

List 2: Messages Exchanged During CRA Re-Authentication

a. $FAP \rightarrow MN : EAP_{req}[ID]$ Inline Formula

b. $MN \rightarrow FAP : EAP_{res}[KeyNameNAI, Seq\#, MAC]$ Inline Formula

c. $FAP \rightarrow FAS : ACC_{req}[KeyNameNAI, Seq\#, MAC]$ Inline Formula

d. $FAS \rightarrow FAP : ACC_{res}[MSK_{CRA}, ReID, EAP_{succes}]$ Inline Formula

e. $FAP \rightarrow MN : EAP_{req}[ReID, Seq\#, MAC]$ Inline Formula

f. $MN \rightarrow FAS : EAP_{res}[ACK, Seq\#, MAC]$ Inline Formula

g. $FAS \rightarrow MN : EAP_{suc}$ Inline Formula

Upon receiving the message the FAS checks the Kname-NAI with its stored authentication information. If there is a match, the server generates the hash value to verify the validity of the message and update the CRA_counter and CRA_timer values. The FAS will then send MSK, MAC, SEQ number to the authenticator. The authenticator retains the MSK and sends the rest to the peer. In the final step, the peer sends an EAP-Response as an acknowledgment. At this point the client is able to calculate the keying material, however to start secure communication the peer waits until it received the EAP-success from the authenticator.

Two sequence numbers, one with HAS and other with FAS are maintained for replay protection of EAP-CRA messages. The sequence number maintained by the supplicant and HAS is initialized to zero on generation of EMSK. The server sets the expected sequence number to the received sequence number plus one on every successful Re-authentication request, i.e. on generation of DSRK. Similarly, the supplicant and the FAS maintain a sequence number with the generation of rMSK while the supplicant is in the FAS's domain.

4.3. Analysis

To substantiate the effectiveness our protocol we first examine the key security features of Enhanced CRA and then compare the cost involved in communication and computing between Enhanced EAP-CRA and its close competitor EAP-ERP.

4.3.1. Security consideration

RFC-3748 [17] indicates mandatory properties and security constraints of an EAP method such as freshness of session key and resistance against replay, dictionary and man in middle attacks. These features can be used as a reference to analyze the protocol in compliance with the EAP frame work. In this section we present our analysis of our protocol against this criterion.

Replay attacks: Generally replay attacks are initiated by re-using captured PDUs. The captured PDUs have authentic ingredients and can be replayed influencing legitimate nodes to respond. The CRA responds to this threat by the use of sequence numbers that enables both the sender and the receiver to have a record of the received datagram. If a packet is out of order it can be dropped. In case of re-authentication the sequence number is generated by the peer. For the rest of the session the peer and the foreign server will increment the value of this sequence number. In the process of full authentication the peer and HAS can benefit from the same procedure to protect against reply attacks.

Man In The Middle (MitM) attacks: In this category of attacks a rogue node introduces itself as a legitimate member in the communication. If there is no security mechanism in place the malicious node can continue to remain in between two legitimate nodes and subsequently masquerade as a legitimate node. During the EAP-CRA re-authentication process, MitM attacks are shunned with a Message Authentication Code (MAC). The MAC is simply a hash of the entire message that is attached to the original message. In this situation an attacker needs to have the knowledge of the hash key to revise the message and to re-calculate the hash. In case of full authentication, the use PKI certificate provides immunization against modification of messages.

Hiding User identification: The proposed method uses KeyName value as user's id during the full CRA process. This prevents from the real identity being revealed to an outsider. During the full authentication process, just before the EAP-Success message the FAS pass a re-authentication ID to the Peer in a secured message. Therefore when the peer requests for re-authentication there is a new random identifier for the peer.

Mutual Authentication:One of the essential features of every EAP method is mutual authentication. However, at the time of publishing EAP framework, the scope of EAP authentication was limited to peer-to-server authentication and the roaming attribute had not been considered. EAP-ERP may satisfy the condition of mutual authentication between Home server and the supplicant, but it is lacking of bilateral proof of identity between the supplicant and a foreign server. More importantly it relies on the security of RADIUS for server-to-server authentication. In contrast, EAP-CRA reaps the advantages of PKI to satisfy this need during the full authentication process.

As both EAP-CRA and EAP-ERP extend the scope of authentication process, the mutual authentication issue can be explored in three areas; between peer and home server, peer and foreign server, and the foreign and home servers. During full EAP-CRA authentication, the proof of possession of MSK (or a key generated from MSK) from the prior EAP authentication process validates the mutual identity between the peer and the home server. The mutual identity between the peer and the foreign server is realized by the foreign server generating a MAC from a key derived from the EMSK which both the foreign server and the peer are in possession. In return the peer also calculates a MAC value to place it inside the final message. This same model is valid for re-authentication phase as well.

Mutual authentication between servers is realized by each server using its private key to encrypt their hostnames. In this view, both servers sign the MSKname to authenticate each other.

4.4. Cost consideration

In this section we compare the cost of communication and computation between Enhanced EAP-CRA and EAP-ERP. It should be noted that EAP-ERP performs a full authentication with the home server every time it enters a foreign network. For this purpose we use EAP-TLS as the home authentication method.

EAP-CRA exchanges eight messages between the supplicant and the servers during full authentication. It also utilizes seven messages during the re-authentication process. In the case of ERP, a minimum of sixteen messages are exchanged between the supplicant and the servers.

This is made up of seven messages that are specific to ERP and at least nine messages from EAP-TLS, since we consider EAP-TLS as the home authentication method. For simplicity we are considering the size of the messages during these exchanges. Table 3 lists the number of messages used in each authentication methods.

When entering a foreign network, a station that uses EAP-ERP performs a full authentication with its home server. This process will be very time consuming due to the fact that all message exchanges should take place over the internet. This is a significant weakness of EAP-ERP compared to EAP-CRA for two reasons; 1) the number of messages and 2) the size of the messages. With regards to re-authentication, ERP re-authentication should take place much quicker as it uses only five messages. However, the actual time differences must be determined after the real setup of both protocols.

Authentication Method	No. of Messages
CRA Full Authentication	7
CRA Re Authentication	8
ERP Initial	16
ERP Re Authentication	5

Table 6. Communication Cost.

To evaluate the computational cost of the protocols we investigate the number of Hashing, Encryption and Decryption operations performed. Table 6 presents these values for EAP-CRA and EAP-ERP. In case of EAP-CRA full authentication there are four hashing operations and eight encryption operations. Initial EAP-ERP does not involve any encryption or decryption but it should be noticed that there will be at least 16 message exchanged while there are just 8 messages for full EAP-CRA authentication. Moreover the encryption involved in the process will ensure the security of the supplicant while it is roaming to a foreign network. In case of Re-authentication, cost of both protocols will be very similar as they both will perform four hash operations.

From the above comparisons we can say that EAP-ERP has high communication costs and Enhanced EAP-CRA has high computing costs. Therefore, we are expecting reasonable performance for Enhanced EAP-CRA due to the fact that communication overheads are normally more costly compared to the computational overheads.

	CRA Full-auth	CRARe-auth	ERPInitial	EAPRe-auth
Sup	Hash(2)	Hash(2)	Hash(0)	Hash(2)
	Encrypt(1)	Encrypt(0)	Encrypt(0)	Encrypt(0)
	Decrypt(1)	Decrypt(0)	Decrypt(0)	Decrypt(0)
FS	Hash(2)	Hash(2)	Hash(0)	Hash(2)
	Encrypt(1)	Encrypt(0)	Encrypt(0)	Encrypt(0)
	Decrypt(1)	Decrypt(0)	Decrypt(0)	Decrypt(0)
HS	Hash(0)	Hash(0)	Hash(0)	Hash(0)
	Encrypt(2)	Encrypt(0)	Encrypt(0)	Encrypt(0)
	Decrypt(2)	Decrypt(0)	Decrypt(0)	Decrypt(0)

Table 7. Computational Cost

5. Conclusion

The main advantage of the CRA mechanism is the use of only two messages to authenticate a wireless device in a FOREIGN network. Although the time taken between the FAS and the HAS may vary depending on the traffic and/or capacity of the wired network, the use of only two messages in a FOREIGN network makes the CRA mechanism very much reliable compared to other available techniques. Further, even if the foreign network uses a less secure authentication mechanism, it still will not affect the CRA clients since their MSKs are supplied by the HASs not-withstanding the limitations of the foreign network.

Another significant advantage of the CRA is its reliance on the HOME security credentials to secure its clients in the foreign network. Hence, it can be assured that the CRA clients will have the same security guarantee as in their home network in a foreign network. Further, in the case of EAP-TLS authentication with CA-signed PKI certificates, clients will need only one certificate signed by the CA and accepted by the HAS. There will be no need for clients to carry a number of different certificates to authenticate with different networks. Hence, in this context, the CRA facilitates EAP-TLS authentication and makes it more practical and viable.

Although there are many other techniques proposed for coordinated authentication, the triumph of the CRA technique is its simplicity, robustness and versatility. Unlike many other systems that require additional components such as a token management system or the Kerberos servers, the CRA depends only on the existing infrastructure, hence, assuring simplicity. The use of existing CA-signed PKI certificates without necessitating other authentication mechanisms such as tokens or smart cards enables the CRA mechanism to be confined. Further, the CRA mechanism is not limited to WLAN, WiMAX or 4G LTE, it can be effectively used with any wireless network, harnessing the unique security features of that particular wireless network. Furthermore, the authentication mechanism (EAP-TLS, EAP-TTLS, EAP-PEAP etc.) used by the wireless network does not influence the CRA mechanism because it does use any form of mappings between these protocols.

On the negative aspect, the effectiveness of the CRA mechanism will depend on the mutual trust established between the participating AAA servers. If the AAA servers do not have any form of prior agreement, it will be up to the discretion of FAS whether to accept or deny a CRA request.

Author details

E. Sithirasenan, K. Ramezani, S. Kumar and V. Muthukkumarasamy

School of Information and Communication Technology Griffith University, Gold Coast, Australia

References

[1] IEEE StdWireless LAN Medium Access Control (MAC) and Physical Layer (PHY) Specifications", (1999).

[2] IEEE Std(2004). IEEE Standard for Local and metropolitan area networks: Part 19: Air Interface for Fixed broadband wireless access systems., 16-2004.

[3] Ghosh, A, Ratasuk, R, Mondal, B, Mangalvedhe, B. N, & Thomas, N. T. LTE-advanced: next-generation wireless broadband technology", in *IEEE Wireless Communications,* Aug. (2010)., 17(3), 10-12.

[4] He, C, & Mitchell, J. C. Security Analysis and Improvements for IEEE 802.11i", in *Proceedings of the 12th Annual Network and Distributed System Security Symposium,* NDSS (2005)., 90-110.

[5] Perrig, A, Stankovic, J, & Wagner, D. Security in wireless sensor networks", Wireless Personal Communications, (2006)., 37(3-4)

[6] IEEE Standard 802i Part 11, "Wireless Medium Access Control (MAC) and Physical Layer (PHY) specifications. Amendment 6: Wireless Medium Access Control (MAC) Security Enhancements," July (2004).

[7] Lynn, M, & Baird, R. Advanced 802.11 attack, Black Hat Briefings, July (2002).

[8] Asokan, N, Niemi, V, & Nyberg, K. Man-in-the-Middle in tunneled authentication protocols. Technical Report (2002). IACR ePrint archive, United Kingdom, Cotober 2002.

[9] IEEE Std 802X-2001, "Local and Metropolitan Area Networks- Port-Based Network Access Control", June (2001).

[10] Iyer, A. P, & Iyer, J. Handling mobility across WiFi and WiMAX", in *Proceedings of the 2009 international Conference on Wireless Communications and Mobile Computing: Connecting the World Wirelessly*, IWCMC (2009). , 537-541.

[11] Machiraju, S, Chen, H, & Bolot, J. Distributed authentication for low-cost wireless networks", in *Proceedings of the 9th Workshop on Mobile Computing Systems and Applications*, HotMobile (2008). , 55-59.

[12] Almus, H, Brose, E, Rebensburg, K, & Kerberos-based, A. EAP method for re-authentication with integrated support for fast handover and IP mobility in wireless LANs", in *Proceedings of the 2nd international conference on communications and electronics*, ICCE (2008). , 61-66.

[13] Huang, Y. L, Lu, P. H, Tygar, J. D, & Joseph, A. D. OSNP: Secure Wireless Authentication Protocol using one-time key", in *Proceedings of Computer and Security* (2009). , 803-815.

[14] Narayanan, V, & Dondeti, L. EAP Extensions for EAP Re- authentication Protocol (ERP)," RFC 5296, Internet Eng. Task Force, (2008).

[15] Salowey, J, Dondeti, L, Narayanan, V, & Nakhjiri, M. Specification for the Derivation of Root Keys from an Extended Master Session Key (EMSK)," RFC 5295, Internet Eng. Task Force, (2008).

[16] Sithirasenan, E, Kumar, S, Ramezani, K, & Muthukkumarasamy, V. An EAP Framework For Unified Authentication in Wireless Networks". In TrustCom'11: *Proceedings of the 10th IEEE International Conference on Trust, Security and Privacy in Computing and Communications*, Nov. (2011). , 92-99.

[17] Blunk, L, & Vollbrecht, J. PPP Extensible Authentication Protocol (EAP)," RFC 3748, Internet Eng. Task Force, (2004).

[18] Stanke, M, & Sikic, M. (2008). *Comparison of the RADIUS and Diameter protocols.* Paper presented at the Information Technology Interfaces, 2008. ITI 2008. 30th International Conference.

[19] Aboba, B, & Simon, D. PPP EAP TLS Authentication Protocol," http://tools.ietf.orgwg/pppext/draft-ietf-pppext-eaptls/draftietf-pppext-eaptls-06.txt, August (1999).

Chapter 6

Key Management in Mobile WiMAX Networks

Mohammad-Mehdi Gilanian-Sadeghi,
Borhanuddin Mohd Ali and Jamalul-Lail Ab Manan

Additional information is available at the end of the chapter

http://dx.doi.org/10.5772/56154

1. Introduction

Wireless networks, because of their many advantages in comparison with the wired ones, have become the predominant technology for deployment of communications infrastructure. WiMAX (Worldwide Interoperability for Microwave Access), which is an industry branding for IEEE 802.16 Wireless Metropolitan Area Network (MAN) sets of standard [1, 2], provides wireless access to mobile devices with a range of Quality of Service (QoS) guarantees for various types of applications. There are diverse versions of IEEE 802.16 standard, but IEEE 802.16e [3] also known as Mobile WiMAX, is the most well-known version, though newer versions have also been formulated.

As for the security model of IEEE 802.16, it has been designed to guarantee authentication, confidentiality and integrity. Among the series of IEEE 802.16 standards, the IEEE 802.16d [4] was defined for fixed wireless access. It uses Privacy Key Management Version 1 (PKMv1) to define, manage and distribute the security keys, but there are several security issues in PKMv1. Hence, in IEEE 802.16e, an enhanced key management scheme called Privacy Key Management Version 2 (PKMv2) was introduced to mitigate the security shortcomings of PKMv1. The PKMv2 uses Extensible Authentication Protocol (EAP) [5] and RSA algorithm [6] as authentication methods. The authentication mechanism ensures that when a Mobile Station (MS) enters a Base Station (BS) coverage area, it should perform authentication and authorization in order to obtain the security keys that will protect data more securely.

The rest of this chapter is organized as follows. Section 2 elaborates the main concept of WiMAX architecture which focuses mainly on the security parts. Section 3 reviews key management protocols in Mobile WiMAX. Finally, section 4 presents our conclusion and suggestions for future works.

2. WiMAX architecture

2.1. Protocol stack

The protocol stack of IEEE 802.16 standard consists of two main layers: Medium Access Control (MAC) layer and Physical (PHY) layer [2]. The MAC layer is subdivided into three sub-layers [7], namely it (CS), Common Part Sub-layer (CPS) and Security Sub-layer (SS) as shown in Figure 1.

The service specific convergence sub-layer communicates with higher layers and receives packets from them and then do some specific functions like packet/frame classification and header suppression. Next, it encapsulates these packets into MAC Service Data Unit (MAC SDU) format, and then distributes MAC SDUs to common part sub-layer. Asynchronous Transfer Mode (ATM) convergence and packet convergence sub-layers are two types of service specific convergence sub-layer. The ATM convergence sub-layer is used for ATM networks, and the packet convergence sub-layer is used for packet services like Ethernet, IPv4 and IPv6.

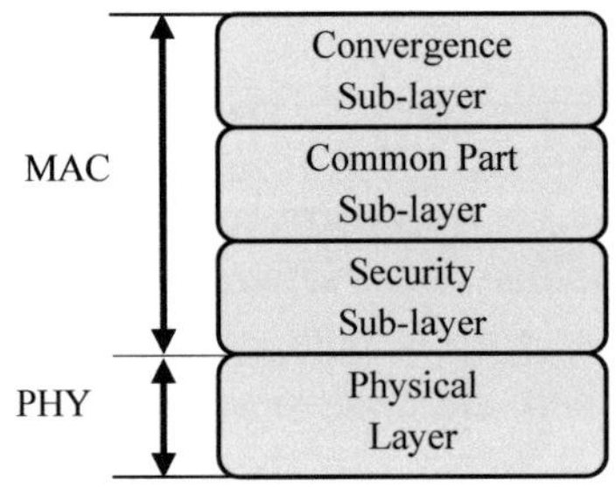

Figure 1. Protocol stack of IEEE 802.16 [2]

The main part of the IEEE 802.16 standard is common part sub-layer which is responsible for bandwidth allocation, connection management, scheduling, connection control, automatic repeat request and QoS enforcement.

The security sub-layer is responsible for providing authentication, authorization and secured key exchange. It is also used for encryption and decryption of data from the MAC layer to PHY layer and vice versa. Two main protocols of security sub-layer are [3]:

1. Encapsulation Protocol, which is used for ciphering operations on data in the networks,

2. PKM protocol, which is used for secure key distribution between BS and MSs, and also it enables the BS to enforce conditional access to network services.

The PHY layer receives MAC frames and then transmits them through coding and modulation of radio frequency signals. It supports Frequency Division Duplexing (FDD) and Time Division Multiplexing (TDM).

2.2. Security sub-layer

The architecture of security sub-layer is shown in Figure 2. As mentioned previously, the security sub-layer provides security services for the standard, and it has been made based on two main components; an encapsulation protocol and a key management protocol [3]. The encapsulation protocol introduces the encryption and authentication methods as cryptographic suites which is a pair of encryption and authentication algorithms.

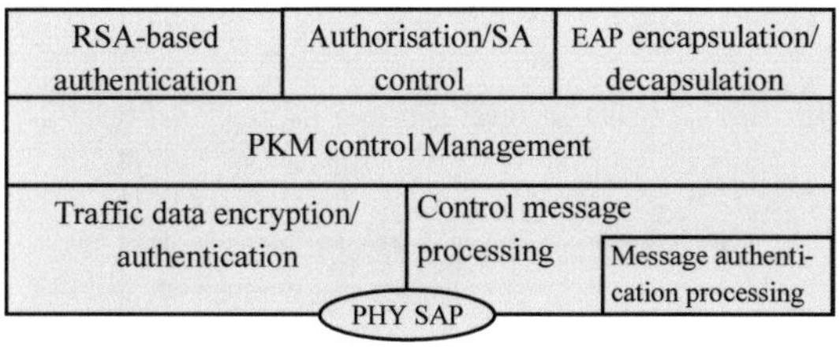

Figure 2. Security sub-layer architecture [3]

Initially, WiMAX security was introduced in the security sub-layer of IEEE 802.16 standard [1]. After releasing the initial versions of the IEEE 802.16 standard, a number of articles such as in [8-10] criticized the security weaknesses of the standard, after which some security improvements were added in IEEE 802.16e [3] and IEEE 802.16m [11]. The security functions regarding key managements have been addressed by PKM protocol. In IEEE 802.16d [4], the key management is based on PKMv1 while IEEE 802.16e uses PKMv2, which is an enhanced version of PKMv1.

Generally, PKM protocol is responsible for authorization, authentication, key exchange and data encryption in the networks between the MSs and BS. In the subsequent sections, we focus our attention on PKMv2, because it is stronger than PKMv1 in terms of security. Recently, the PKMv3 [11] was launched with IEEE 802.16m standard, however, since this protocol is still new and only a few works are being done on it, it is not discussed further in this chapter.

The PKMv2 is used by MSs to get authorization and security keys from the BS, and also to guarantee continuous and uninterrupted re-authorization/re-authentication and refreshing of the security keys. The PKMv2 applies EAP protocol together with RSA algorithm or a mixed function starting with RSA followed by EAP. As shown in Figure 3, in EAP of PKMv2, the root of the security keys is Master Session Key (MSK), and the other keys such as Key Encryption Key (KEK) are obtained from the MSK.

The procedure of security keys generation using the EAP method is shown in Figure 3. In this Figure, the result of EAP authentication protocol is MSK. Then both the MS and BS make a Pairwise Master Key (PMK) by removing some bits of the MSK using a number of functions such as Dot16KDF [12], and also they generate an Authorization Key (AK) from the PMK. After making the AK, the BS and MS will establish the Key Encryption Key (KEK) from the AK. The BS and MS use a 3-way handshake to drive Traffic Encryption Key (TEK) which is used to encrypt data in the network between the BS and MSs. The Multicast Broadcast Service (MBS) is then applied when there are several MSs whereby the MBS is used to send the messages to them. In this case, both BS and MS need to generate and use some group keys.

IEEE 802.16 supports multicast for applications such as pay-tv and videoconferencing. In order to establish a secure multicast over IEEE 802.16, main components of the standard must be used, namely Multicast Broadcast Service (MBS) and Multicast and Broadcast Rekeying Algorithm (MBRA). We will explain how this is done in the next section.

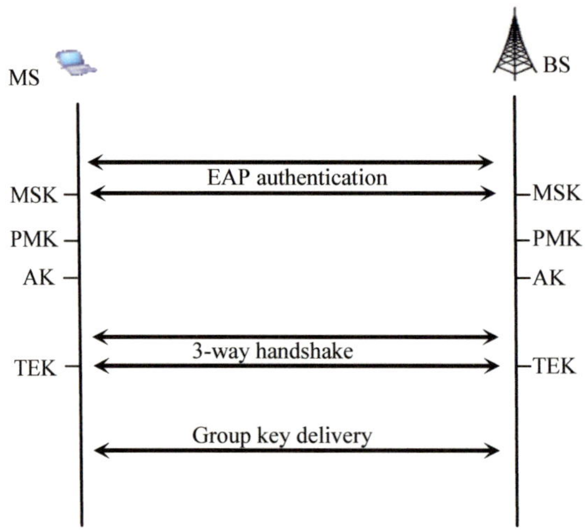

Figure 3. Key generation at initial network entry [12]

3. Key management protocols

There are three types of Key management protocols, viz, centralized, hierarchical and distributed key management [13]. WiMAX Network uses a centralized key management where there is a single manager (BS) that executes key management procedure for all its members (MSs). Though some key management protocols have been proposed for WiMAX, their protocols still remained inefficient.

Generally, key management establishes a set of group keys for its members [14], and the main function of it is to update the group keys, this is called rekeying algorithm [15]. The key management protocols have to face several challenges, but the most outstanding challenges among them are on performance and security, as shown in Figure 4. Under performance are issues such as operational efficiency, scalability and 1-affects-n phenomenon [16, 17]. Operational efficiency is the most important parameter in performance measure and is measured typically by storage, communications and computational costs respectively. In measuring the performance of key management, the storage costs refer to the number of keys stored by the BS and MSs; the communications costs refer to the number of transmitted group keys upon a rekeying algorithm, and the computational costs refer to the cost of ciphering operations in order to get the updated group keys. Scalability means the capability of key management protocol to handle a large group of members, and also its ability to manage highly dynamic membership changes. The 1-affects-n phenomenon is estimated from the number of members affected by rekeying operations. Moreover, a key management should support forward secrecy, which means that the MSs that leave a BS cannot read future messages; and also it must guarantee backward secrecy, which means that a new MS cannot read previous messages [9].

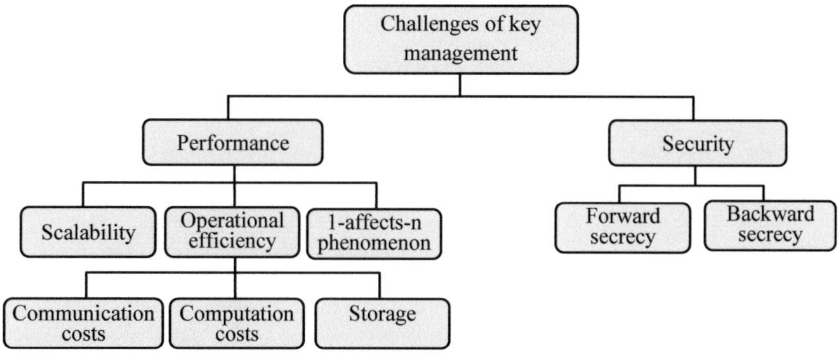

Figure 4. Key management's challenges

3.1. Multicast and broadcast service

Multicast and Broadcast Service (MBS) of IEEE 802.16e is a new feature for broadband wireless standards [3]. It is a mechanism that allows a BS to distribute the same set of data to several MSs concurrently. As highlighted before, first the MSs need to be authenticated by the BS using PKMv2 [3]. After that, the Group Key Encryption Key (GKEK) and the Group Traffic Encryption Key (GTEK) are established. IEEE 802.16e introduced the MBRA as a basic rekeying algorithm to generate, update and distribute the GKEKs and GTEKs upon member changes. The MBRA uses the GTEK which is shared among all MSs to encapsulate the data traffic. The BS generates the GKEK and the key is used to encapsulate the GTEK. The GKEK is also encapsulated by the KEK of each MS. Each MS has a unique KEK which is obtained from the AK. Although, the MBRA of MBS is quite well designed, it still suffers from efficiency and scalability problem and it does not address backward and forward secrecy [8, 18]. To explain this point, in the MBRA algorithm, the BS should unicast n messages, where n is the number of MS, with the aim of updating the group keys, which unfortunately would cause weak scalability due to the increased number of unicast messages. Moreover, when there are high numbers of MSs, and the effect of sending high volume of unicast/broadcast messages would increase communication costs, and consequently this will result in poor efficiency.

Rekeying algorithms in WiMAX networks need to execute using one of the following three events:

1. when a new MS joins the BS,
2. when the life time of both GTEK and GKEK expire,
3. when an MS leaves the BS.

The MBRA algorithm of Mobile WiMAX which is a simple rekeying will only happen at the expiration time of GTEK or GKEK. As shown in Figure 5, from time to time, the BS broadcasts message (1) encrypted by GKEK to all MSs in order to update the GTEK as well as sending a unicast message (2) to all MSs which has been encapsulated by the KEK of each MS as shown by the equations below:

$$BS \Rightarrow all\ MS : \{GTEK\}_{GKEK} \tag{1}$$

$$BS \rightarrow each\ MS : \{GKEK\}_{KEK} \tag{2}$$

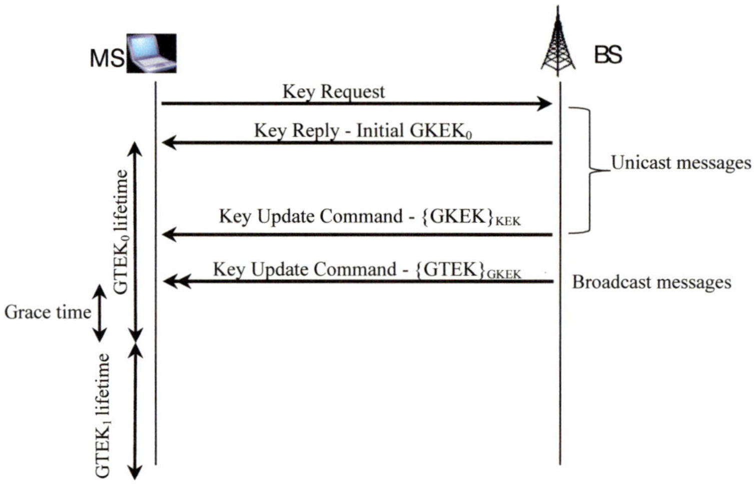

Figure 5. MBRA messages [19]

The nomenclatures are listed as in Table 1.

X⇒Y	X broadcasts a message to Y
X→Y	X unicasts a message to Y
$[X]_Y$	X encrypted by using key Y
MS_{SGi}	The collection of all MSs within $subgroup_i$

Table 1. Nomenclature of key management

3.2. Rekeying algorithms

As mentioned earlier, in the MBRA algorithm, the number of unicast messages on rekeying increase with the number of MS, and hence this method is neither scalable nor efficient. In addition, it does not address forward and backward secrecy, which consequently would lead to this method being vulnerable to attacks [8, 9].

Researchers in [18] performed a detailed analysis of the MBRA algorithm and identified its deficiencies. They proposed an improved scheme to address the deficiencies identified. Even though their method showed some improvements on the MBRA, but they suffer a downside in that, the BS needs to send n (n being the number of MS) unicast messages upon every membership changes, which consequently resulted in the drastic drop in network efficiency for a large number of MSs. In addition, the proposed method also sends some plaintexts for message broadcasting, which could cause critical security breaches [19]. In fact, despite some improvements to the MBRA rekeying, the proposed method suffers from some security issues such as Denial of Service (DoS) [19] as well as poor scalability and efficiency. In addition, it does not address the 1-affects-n phenomenon [17], very well.

The authors in [20] proposed a new group key management protocol called Group-Based Key Distribution Algorithm (GKDA) in which the security keys are distributed into subgroups. The MBS group is first divided into N subgroups; hence, N GKEKs for the subgroups are used instead of one GKEK being shared among all MSs. By doing so, only that GKEK which is used for a certain subgroup needs to be updated whenever any membership change (e.g. leave event) occurs in that subgroup. The GKEK is encapsulated by each MS's KEK in the subgroup, and then unicast to each MS. Although the GKDA provides forward and backward secrecy, it is still not scalable and efficient enough, because when the number of MS in each subgroup grows bigger, the number of unicast messages to update GKEKs grows likewise. Nevertheless, GKDA is still better than MBRA in terms of reducing the number of unicast messages needed to perform updates of the group keys. In GKDA, the GTEK update mode is more lengthy because it consists of N GTEKs which are encapsulated by N GKEKs, and thus it consumes more energy to send the messages. Moreover, the scheme does not have a good support for 1-affects-n phenomenon.

In [21], the authors proposed an algorithm called Efficient sub-Linear rekeying Algorithm with Perfect Secrecy (ELAPSE) in order to address the problems of MBRA algorithm. Although this method solves the forward and backward secrecy problems, it suffers from some weaknesses in terms of scalability and efficiency. In ELAPSE, when member join or leave events happen frequently within a large group, the overall performance will degrade due to communication and computational costs. This method is based on key hierarchy and sub-grouping of the MSs in the cell by means of a binary tree. ELASPE divides the number of MS into N=log (n,2) subgroups where n is the number of MS, and each subgroup keeps a set of hierarchical keys named Sub Group KEKs (SGKEKs) instead of a single GKEK. The number of subgroups (N) is defined in advance by the administrator depending on the application's requirements, i.e. the number of subgroups is permanent. The result is weak performance in terms of efficiency and scalability. We illustrate this issue by way of an example as shown in Figure 6, which shows a binary tree with four subgroups. All MSs maintain similar GTEK, and each MS in each subgroup saves a set of SGKEKs; for example, the MSs in $subgroup_1$ store three group keys $SGKEK_1$, $SGKEK_2$ and $SGKEK_{1234}$. The $SGKEK_{1234}$ is similar with the GKEK in MBRA. In this case, GKEK is not delivered to each MS by unicast message, instead it is distributed among the subgroups via broadcast messages.

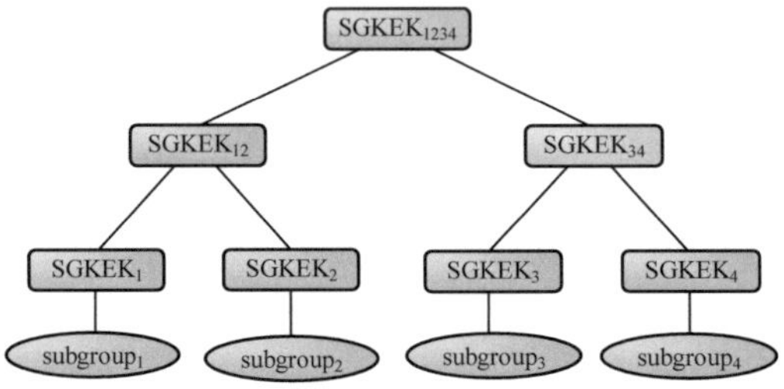

Figure 6. Key Hierarchy with four subgroups [21]

When there is no new member joining or leaving, and the lifetime of GTEK expires, the BS broadcasts a new GTEK encapsulated by $SGKEK_{1234}$ to all MSs represented as message (3) below.

$$BS \Rightarrow all\ MSs : \{GTEK\}_{SGKEK_{1234}} \tag{3}$$

Upon a member join event, i.e. when a new MS enters into the BS coverage area, and $subgroup_2$ has the lowest number of members, then the BS assigns it to $subgroup_2$. The BS unicasts message (4) below to the new MS and all MSs in $subgroup_2$ in order to update the group keys. Message (4) is then encapsulated by KEK of each MS, and contains all new group keys from $subgroup_2$ to the root of binary tree.

$$BS \rightarrow MS_{SG2}\ \&\ new\ MS : \{GTEK, SGKEK_{1234}, SGKEK_{12}, SGKEK_2\}_{KEK} \tag{4}$$

In order to update the group keys as well as to provide the backward secrecy, the BS needs to send two broadcasts i.e. messages (5) and (6) below, to all MSs excluding $subgroup_2$.

$$BS \Rightarrow MS_{SG3}, MS_{SG4} : \{GTEK, SGKEK_{1234}\}_{SGKEK_{34}} \tag{5}$$

$$BS \Rightarrow MS_{SG1} : \{GTEK, SGKEK_{1234}, SGKEK_{12}\}_{SGKEK_1} \tag{6}$$

Upon member leave event, i.e. when a MS leaves the BS coverage area, the process of the group key updating is similar to member join event. For instance, when one MS of $subgroup_2$ leaves the BS, then the BS should unicast message (7) to all remaining MSs in $subgroup_2$. It also needs to broadcast two messages, i.e. messages (8) and (9), to all MSs except $subgroup_2$.

$$BS \rightarrow MS_{SG2} : \{GTEK, SGKEK_{1234}, SGKEK_{12}, SGKEK_{2}\}_{KEK} \tag{7}$$

$$BS \Rightarrow MS_{SG3}, MS_{SG4} : \{GTEK, SGKEK_{1234}\}_{SGKEK_{34}} \tag{8}$$

$$BS \Rightarrow MS_{SG1} : \{GTEK, SGKEK_{1234}, SGKEK_{12}\}_{SGKEK_{1}} \tag{9}$$

Authors in [22] suggested an improved version of ELASPE called ELAPSE+ using cross layering concept. They assigned fast moving MSs such as cars to specific subgroups, and made the size of those specific subgroups to be smaller than the other subgroups. This is because, the fast moving MSs pass through the BS's cell length faster, and therefore they would experience high number join or leave events, which gives rise to the need to update more group keys. Although ELAPSE+ improves the performance of ELAPSE by reducing the amount of rekeying messages needed to send unicast and broadcast messages, it still inherits the drawback of handling static numbers of subgroups, subsequently resulting in weak efficiency and scalability.

The authors in [23] proposed a hybrid key management scheme to improve the performance of ELAPSE and ELASPE+ upon rekeying by reducing message passing. This scheme uses the architecture of LORE [23] within a subgroup of ELAPSE. In this way, when a MS enters a BS coverage area, the BS assigns it to a subgroup and also provides a Subgroup Forward Key Set (SGFSet) and Subgroup Backward Key Set (SGBSet). These key sets are created by simple Pseudo-Random Generator (PRNG) and keep the ordering of nodes inside a subgroup similar to LORE. Hence, if there are k MSs in a subgroup, then there are k numbers of Subgroup Forward Key (SGFK) and k numbers of Subgroup Backward Key (SGBK). In this way, for each MS i there are two sets of keys as follows:

$$SGFSet = \{SGFK_{m} \mid 1 \leq m \leq i\} \tag{10}$$

$$SGBSet = \{SGBK_{m} \mid i \leq m \leq k\} \tag{11}$$

Figure 7 shows the revised version of ELAPSE. Here, a node i in $subgroup_2$ has three keys $SGKEK_{1234}$, $SGKEK_{12}$ and $SGKEK_2$ as well as a two-key set $SGFSet^{i}{}_{2}$ and $SGBSet^{i}{}_{2}$. Upon member join or leave event, the rekeying algorithm updates SGKEKs and GTEK, but there is no change in SGFSet and SGBSet sets. After a predefined time T, both SGFSet and SGBSet will be renewed.

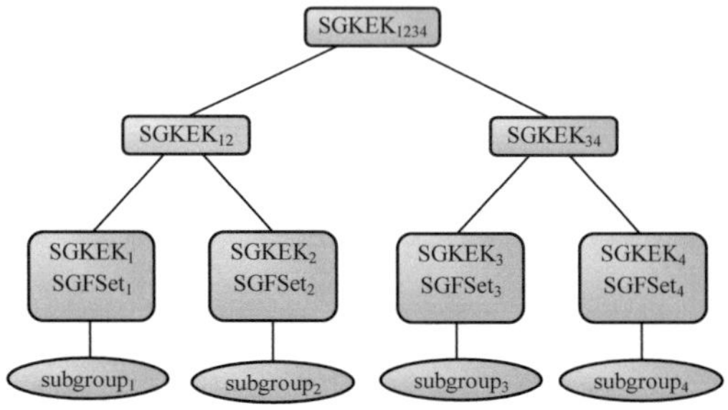

Figure 7. A revised version of ELAPSE [23]

It should be noted here that this improvement in communication costs over ELAPSE, comes at high computational and storage costs in the revised version of ELAPSE. Moreover, this scheme gives rise to security issues such as collusion resistance [23] which means two or more MSs must not get secret keys that they are not allowed to know, and this could be done by exchanging their respective secret keys.

The authors in [24] improved ELAPSE by using a n-ary tree (where n>2) to improve the efficiency of key management. Even though the proposed method shows some improvements on the efficiency of ELAPSE, the method still suffers from the limitations associated with fixed number of subgroups. In this method, the tree depth becomes large when the number of MS increases, and this is the main issue with a binary tree. Therefore, they suggested that by using n-ary, the efficiency of group key updating algorithm will improve. Figure 8 shows a 3-ary tree with 9 subgroups.

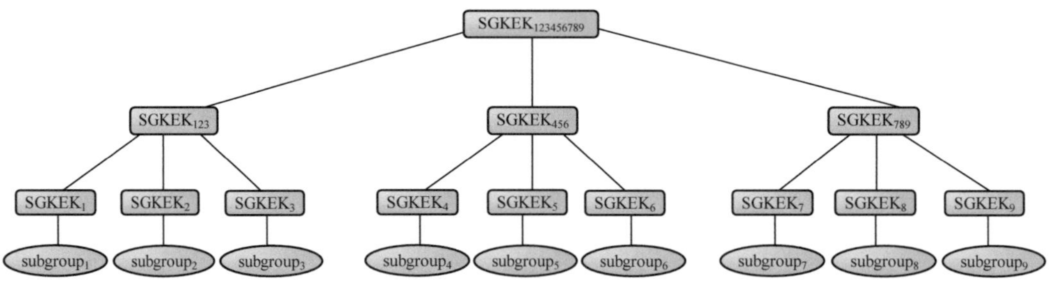

Figure 8. A 3-ary tree [24]

The number of group keys in n-ary tree and the tree depth are given by equation (12) and (13) respectively.

http://dx.doi.org/10.5772/56154

$$k = \sum_{i=0}^{d} n^i \tag{12}$$

$$d = \left\lceil \log_n^N \right\rceil \tag{13}$$

By using n-ary tree, the BS needs to keep more group keys compared with ELAPSE method. So, in terms of storage costs n-ary tree does not perform very well, even though the communication costs is considerably decreased due to the reduction in communication overheads upon group keys updating. The authors made detailed analysis to find the optimal value of n in order to minimize the total energy consumption of the rekeying algorithm. They assumed that transmission and reception energy are equal to total energy consumption of the networks, whereby the energy consumption refers to the length of broadcast or unicast messages. Finally, they came out with an optimal value of n=4, meaning that 4-ary tree would give the best performance in terms of energy consumption.

It should be highlighted here that basically the methods in [20, 21, 23, 24] are based on ELAPSE in that they all use tree structures, and therefore the problem associated with ELAPSE as highlighted before still remains.

The authors in [19] proposed a new method of improving MBRA using asymmetric algorithms. The idea of this method is to establish a common encryption key which is shared among all MSs, but every MS has a different decryption key. This means that the BS can encrypt the messages including the group keys, and only the valid MS can decrypt the messages. In this way, the proposed method provides backward and forward secrecy. In terms of operational efficiency, this method needs to perform more computations because of the use of asymmetric cryptography, and hence this makes the MSs to expense more energy which is not good for mobile devices. Nevertheless, one advantage of the proposed method is that it sends less unicast/broadcast messages, and hence the overall communication cost is low. In this way, upon member changing, the BS sends one broadcast message, but on normal key refresh, it needs to send n unicast messages, where n is the number of MS and also the BS should send two broadcast messages. The proposed method managed to address the backward and forward secrecy issue of the MBRA algorithm. However, it has poor response to scalability, since upon group key updating after the expiration time, it has to send n unicast messages. Moreover, the method needs to make numerous modifications to the standard, which it is not practical for implementation in real environment.

In [25] the Scalable Rekeying Algorithm (SRA) is proposed, which is based on complete binary tree [26], and is implemented by linear linked list data structure. The SRA method improves the scalability for ELAPSE and it can also improve the other methods [20, 21, 23, 24] which have similar setups. As mentioned earlier, ELAPSE divides the MSs into N subgroups. In this way, each subgroup keeps a set of group keys. In fact, ELAPSE employs a fixed number of subgroups, consequently upon group key updating, the ELAPSE shows poor scalability. In addition, the method consumes more bandwidth because of the sending of high number of unicast messages.

The SRA method establishes the number of subgroups according to the number of current MS in the cell. Figure 9 shows a sample of node within linear linked list, where "#MS" field indicates all MSs in a certain subgroup. The group key for that subgroup is "Group-key". L1 and L2 are two pointer fields in the node, where L1 points to the MSs of that subgroup and L2 points to the next node (subgroup).

Figure 9. A node of linear linked list

The SRA method uses log (n,2) in order to subgroup the MSs, and whereby according to the current number of MS, it increases or decreases the number of subgroups.

As highlighted before, in Mobile WiMAX, group key updating happens on three events:

1. Upon the expiry lifetime of GTEK/GKEK,
2. Upon member join event,
3. Upon member leave event.

For the first event (i.e. upon the lifetime of GTEK or GKEK expiry), the SRA and ELAPSE methods apply similar functions. However, in SRA method, on member join/ leave, it is necessary to add/delete a subgroup at a certain time to increase or decrease the number of subgroups based on log(n,2).

Assuming that in the first step there is one subgroup as shown in Figure 10, it means that there is a node in the linear linked list. Figure 10 shows a linear linked list structure corresponding to complete binary tree. For the rest of the chapter, the tree is not drawn for the sake of simplicity.

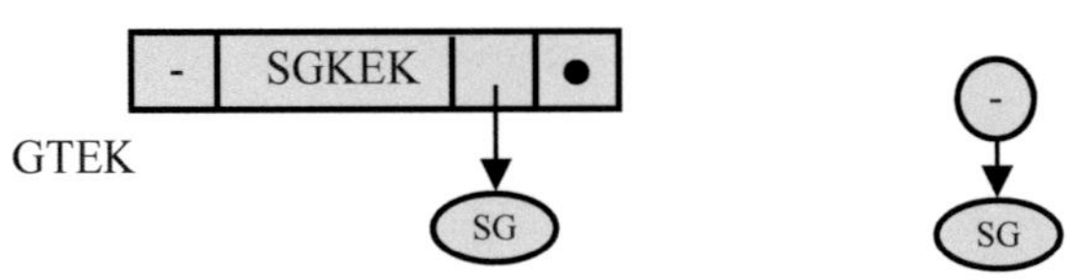

(a) Linear linked list structure

(b) Complete binary tree structure

Figure 10. The creation of one subgroup

As the number of MS reaches three, a new subgroup should be added, based on log(3,2)=2. Thus, subgroup SG breaks into 2 subgroups, SG_1 and SG_2. Subsequently, the MSs of SG partition into two different sets, and afterward they are inserted separately into 2 subgroups,

SG_1 and SG_2. In the properties of complete binary tree, if a node is at an index i, the left child is at index 2*i, and the right child is at index 2*i+1. We use these properties of the tree to manage the subgroups. In this way, SG_1 is at index 2 and SG_2 is at index 3. In Figure 11, two subgroups are shown with 1 and 2 MSs in '#MS' field respectively.

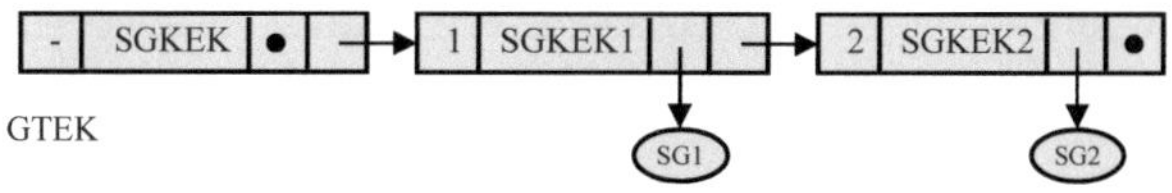

Figure 11. The creation of two subgroups

The BS unicasts two messages i.e. messages (14) and (15) to all MSs with the purpose of updating the group keys. In this way, the BS unicasts $SGKEK_1$ and $SGKEK_2$ to SG_1 and SG_2 respectively.

$$BS \rightarrow MS_{SG1} : \{GTEK, SGKEK_1\}_{KEK} \tag{14}$$

$$BS \rightarrow MS_{SG2} : \{GTEK, SGKEK_2\}_{KEK} \tag{15}$$

As the number of MS increases beyond 5, a new subgroup is added, based on log(n,2). In this case, the new node is added to the left side of the tree; the left hand side's children of the tree are regarded as 2 new subgroups. Next, the MSs of SG_1 divides into 2 parts and then they are associated to 2 new subgroups, viz, SG_{11} and SG_{12} as shown in Figure 12.

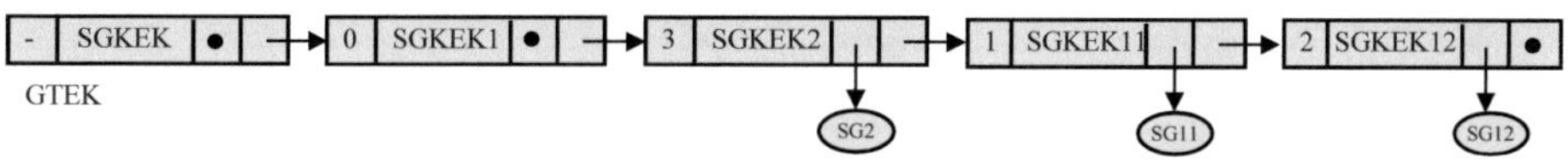

Figure 12. The creation of three subgroups

To update the group keys after inserting one new subgroup, the BS should unicast two messages i.e. messages (16) and (17) to SG_{11} and SG_{12} respectively. Assuming that the BS adds the new MS to SG_{11}, then the new group keys should be unicast to the MS by means of message (16).

$$BS \rightarrow MS_{SG11} \;\&\; newMS : \{GTEK, SGKEK, SGKEK_1, SGKEK_{11}\}_{KEK} \quad (16)$$

$$BS \rightarrow MS_{SG12} : \{GTEK, SGKEK, SGKEK_1, SGKEK_{12}\}_{KEK} \quad (17)$$

If the number of MS is 11 or less, they join three subgroups (Figure 12), but if they exceed 11, one new subgroup must be added. The procedure to add a new subgroup is similar to our explanation for Figure 12. Here, SG_2 divides into two subgroups i.e. SG_{21} and SG_{22}. The entire number of MSs in each subgroup is labeled in '#MS' field of Figure 13, when the 12th MS enters into the BS coverage area.

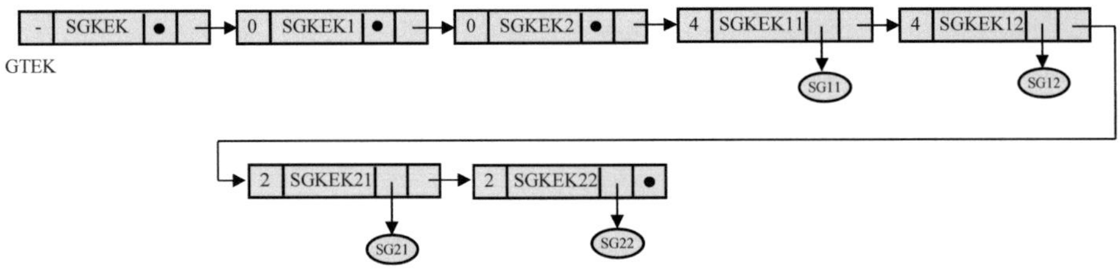

Figure 13. Linear linked list showing the creation of four subgroups

Suppose a few MSs leave a cell, the total number of MS will decrease. As a result, the number of subgroups based on log(n,2) should decrease as well. When the number of MS drops to less than 12, SG_{21} and SG_{22} combine together into one subgroup, i.e. SG_{2b}. Next, the whole MSs in SG_{21} and SG_{22} add into SG_{2b}. When the number of MS stands at 11 the subgroups that exist becomes as shown in Figure 14. The BS unicasts message (18) including 3 new group keys to every MS in SG_{2b} to update the group keys.

$$BS \rightarrow MS_{SG2b} : \{GTEK, SGKEK, SGKEK_{2b}\}_{KEK} \quad (18)$$

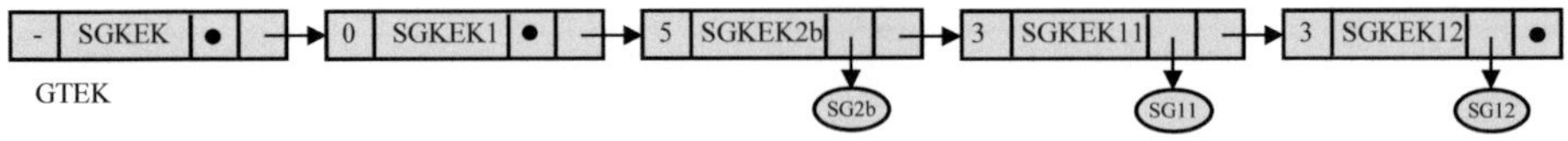

Figure 14. Linear linked list showing the creation of three subgroups

In the forthcoming, the SRA method is compared and analyzed against MBRA [3] and ELAPSE [21]. The MBRA unicasts n messages to all current MSs as well as new MS upon member joining, and upon member leaving, it unicasts n-1 messages (since 1 MS leaves the cell). As mentioned earlier, ELAPSE creates a permanent number of subgroups, therefore when the number of MS in a cell grows, the number of transmitted unicast messages increases likewise. The entire number of transmitted unicast messages in ELAPSE is (n/N); in fact, it is (n/N)+1 for member joining and (n/N)-1 for member leaving. In SRA, the number of MS in each subgroup is n/log(n,2), and therefore the number of unicast messages is likewise n/(log(n,2)) on member joining/leaving. The comparison among the MBRA, ELAPSE and SAR is shown in Table 2, where ELAPSE4 means four subgroups, and ELAPSE8 means 8 subgroups.

Methods	Unicast Messages	
	Join	Leave
MBRA	$O(n)+1$	$O(n)-1$
ELAPSE4	$O\left(n/_4\right)+1$	$O\left(n/_4\right)-1$
ELAPSE8	$O\left(n/_8\right)+1$	$O\left(n/_8\right)-1$
SAR	$O\left(\frac{n}{\log_2^n}\right)+1$	$O\left(\frac{n}{\log_2^n}\right)-1$

Table 2. Comparative analysis [25]

Figure 15a & b show the comparison among the rekeying algorithms in terms of unicast messages. Here, y-axis represents the number of unicast messages, and x-axis is the number of MS. As shown in Figure 15a, in the MBRA, the number of unicast messages increase with growing number of MS, and clearly it does not address the question of scalability. Figure 15b shows the analysis among the tree-based rekeying algorithms only. This is also the case with ELAPSE, where the number of unicast messages increases with the number of MS in each subgroup. On the other hand, in SRA the number of unicast messages is less than in ELAPSE, which means that it provides a good scalability even at high number of MS.

Figure 16 shows a magnified view of Figure 15, for the number of MS between 200 to 400 in Figure 16a, and 500 to 700 in Figure 16b, respectively. It is clear from the Figure that in SRA method, as the number of MS increases the number of transmitted unicast messages increases with a much lesser degree than for ELAPSE. In other words, the difference between the number of unicast messages between SRA and ELAPSE widens. For example, when there are 400 MSs (Figure 16a), the difference between the number of transmitted unicast messages in the SRA and ELAPSE8 is around 5, but when there are 700 MSs (Figure 16b), this difference is around 10. This shows that SRA method has a good scalability performance especially at high number of MS in the cell.

Figure 17 depicts a summarised comparison between SRA and ELAPSE. Clearly, SRA reduces the number of unicast messages upon implementing rekeying algorithm, and therefore it has better scalability compared with ELAPSE. Even though ELAPSE8 shows comparable performance with SRA especially at lower number of MS, the number of MS in a subgroup has to be defined in advance and neither it is dynamic.

Finally, Table 3 summarizes the main characteristics of the rekeying algorithms which have been highlighted in this chapter.

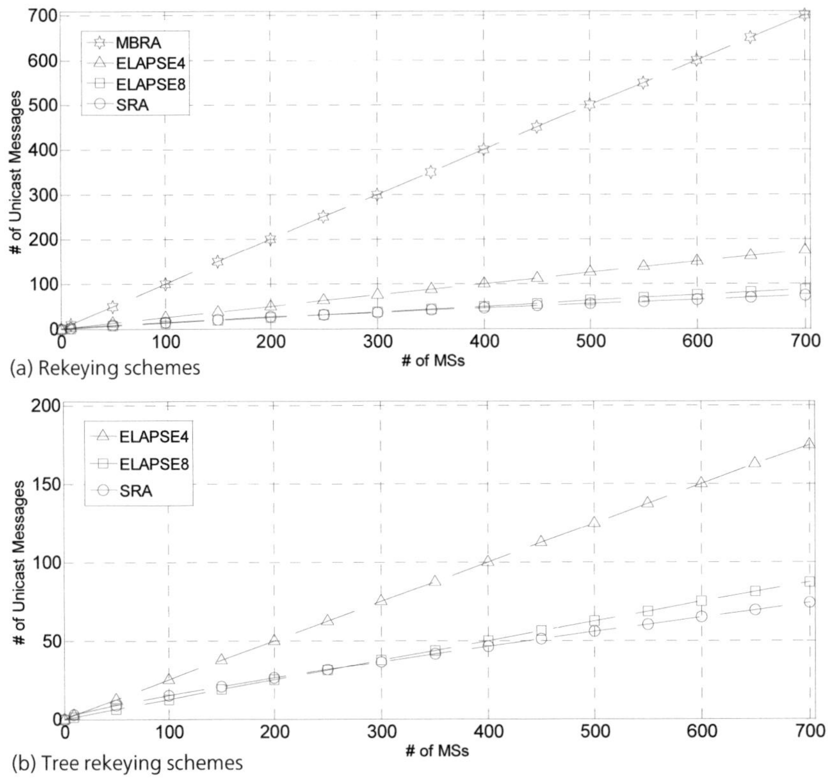

(a) Rekeying schemes

(b) Tree rekeying schemes

Figure 15. Unicast messages

Scheme	*Forward/Backward Secrecy*	*Scalability & 1-affects-n*	Operational efficiency*
MBRA[3]	not supported	Very weak	Non optimal
Xu et al.[18]	Supported	weak	Non optimal
GKDA[20]	supported	good	Non optimal
Chakraborty et al.[23]	supported	good	Non optimal
Kambourakis et al.[19]	supported	good	Non optimal
ELAPSE [21]	supported	good	Non optimal
Brown et al.[24]	supported	good	Non optimal
SRA[25]	supported	Very good	Near optimal

*This shows the trade-off among communication, computational overheads.

Table 3. Summary of the main performance parameters of rekeying algorithms

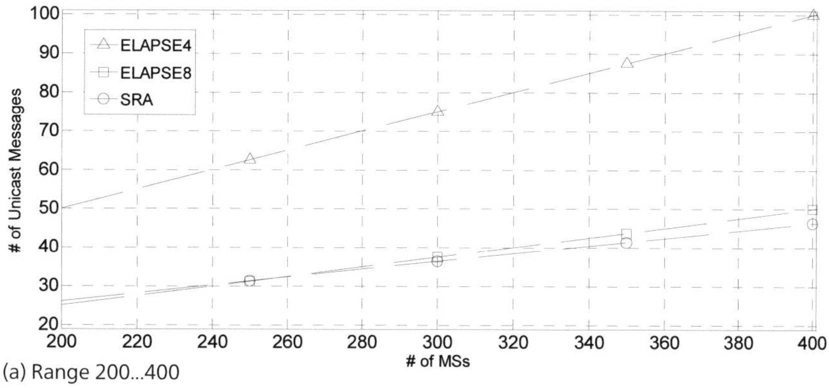

(a) Range 200...400

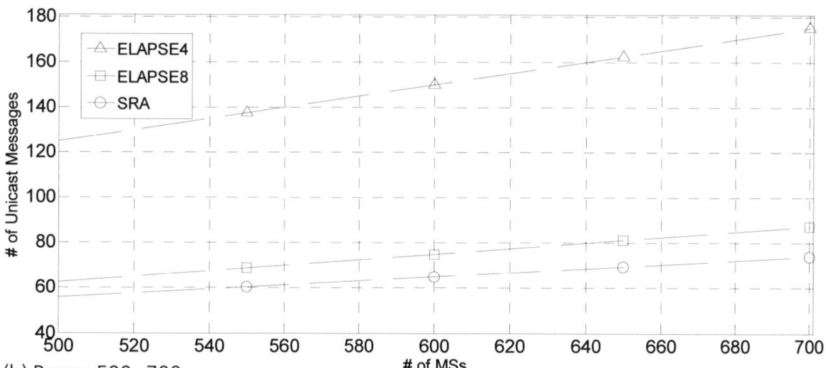

(b) Range 500...700

Figure 16. Unicast messages in tree rekeying

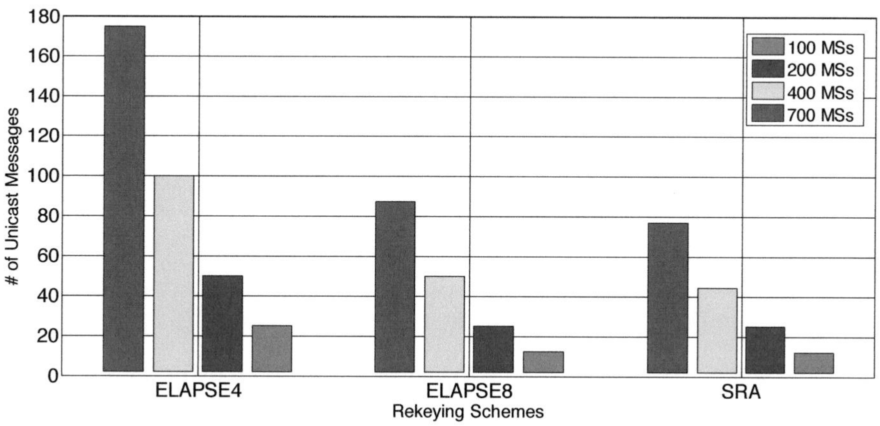

Figure 17. Unicast messages in ELAPSE and SRA

4. Conclusions

In this chapter, we reviewed the MBRA rekeying algorithm of the IEEE 802.16e and analyzed several rekeying algorithms for Mobile WiMAX. We reviewed the rekeying algorithms with emphasis on performance and security particularly their effects on operational efficiency, scalability and 1-affects-n phenomenon as well as backward and forward secrecy. We showed that SRA rekeying algorithm is a strong algorithm from the scalability aspect, because it establishes the number of subgroups dynamically and hence strikes a good balance between the number of MS in each subgroup and the total number of subgroups. The overall result is a reduction in the number of unicast messages on rekeying which produce better scalability and efficiency. The future work will focus on reducing energy consumption in the MSs upon rekeying, by broadcasting the group keys to only the selected MSs that need them, rather than sending to all MSs.

Acknowledgements

This work is supported by Universiti Putra Malaysia and Ministry of Science, Technology and Innovation under the Science-fund (no. 01-01-04-SF1417).

Author details

Mohammad-Mehdi Gilanian-Sadeghi[1], Borhanuddin Mohd Ali[1] and
Jamalul-Lail Ab Manan[2]

1 Department of Computer and Communication Systems Engineering, Faculty of Engineering, Universiti Putra Malaysia, UPM Serdang, Selangor, Malaysia

2 Strategic Advanced Research, MIMOS Berhad, Malaysia

References

[1] "IEEE Std 802.16, IEEE Standard for Local and metropolitan area networks, Part 16: Air Interface for Broadband Wireless Access Systems ", ed: IEEE Press, 2004.

[2] "IEEE Std 802.16, IEEE Standard for Local and metropolitan area networks, Part 16: Air Interface for Broadband Wireless Access Systems and Revision of IEEE Std 802.16-2004," ed: IEEE Press, 2009.

[3] "IEEE Std 802.16e, IEEE Standard for Local and metropolitan area networks, Part 16: Air Interface for Fixed and Mobile Broadband Wireless Access Systems, Amendment

2: Physical and Medium Access Control Layers for Combined Fixed and Mobile Operation in Licensed Bands and Amendment and Corrigendum to IEEE Std 802.16-2004," ed: IEEE Press, 2006.

[4] "IEEE Standard for Local and Metropolitan Area Networks, Part 16: Air Interface for Fixed and Mobile Broadband Wireless Access Systems," ed: IEEE Press, 2004.

[5] B. Aboba, L. J. Blunk, J. R. Vollbrecht, J. Carlson, and H. Levkowetz, "Extensible Authentication Protocol," RFC 3748, 2004.

[6] R. Laboratories, PKCS #1: RSA Cryptography Standard, 2002.

[7] S. Ahson and M. Ilyas, WiMAX: Standards and Security. CRC Press, Inc. Boca Raton, FL, USA, 2008.

[8] D. Johnston and J. Walker, "Overview of IEEE 802.16 Security," IEEE Security and Privacy, vol. 2, pp. 40-48, 2004.

[9] A. Deininger, S. Kiyomoto, J. Kurihara, and T. Tanaka, "Security Vulnerabilities and Solutions in Mobile WiMAX " IJCSNS International Journal of Computer Science and Network Security vol. 7, pp. 7-15, 2007.

[10] T. Shon, B. Koo, J. H. Park, and H. Chang, "Novel Approaches to Enhance Mobile WiMAX Security," EURASIP Journal on Wireless Communications and Networking, Article ID 926275, 2010.

[11] "P802.16m/D6, IEEE Standard for Local and Metropolitan Area Networks - Part 16: Air Interface for Broadband Wireless Access Systems - Advanced Air Interface," May 2010

[12] J. Hur, H. Shim, P. Kim, H. Yoon, and N.-O. Song, "Security Considerations for Handover Schemes in Mobile WiMAX Networks," in IEEE Wireless Communications and Networking Conference (WCNC), 2008, pp. 2531-2536.

[13] S. Rafaeli and D. Hutchison, "A survey of key management for secure group communication " ACM Computing Surveys, vol. 35, pp. 309-329, 2003.

[14] M. Baugher, R. Canetti, L. R. Dondeti, and F. Lindholm, "Multicast Security (MSEC) Group Key Management Architecture," in RFC 4046, 2005

[15] T. Hardjono and L. R. Dondeti, Multicast And Group Security. USA, 2003.

[16] Y. Challal and H. Seba, "Group Key Management Protocols: A Novel Taxonomy," International Journal of Information Technology, vol. 2, pp. 105-118, 2005.

[17] S. Gharout, A. Bouabdallah, M. Kellil, and Y. Challal, "Key Management With Host Mobility in Dynamic Groups," in International conference on Security of information and networks New York, USA, 2010, pp. 186-193.

[18] S. Xu, C.-T. Huang, and M. M. Matthews, "Secure Multicast in WiMAX," Journal of Networks, vol. 3, pp. 48-57, 2008.

[19] G. Kambourakis, E. Konstantinou, and S. Gritzalis, "Revisiting WiMAX MBS security," Computers and Mathematics with Applications, vol. 60, pp. 217-223, 2010.

[20] H. Li, G. Fan, J. Qiu, and X. Lin, "GKDA: A Group-Based Key Distribution Algorithm for WiMAX MBS Security," Advances in Multimedia Information Processing, LNCS, Springer Verlag, vol. 4261, pp. 310-318, 2006

[21] C. T. Huang and J. M. Chang, "Responding to Security Issues in WiMAX Networks," IEEE IT Professional vol. 10, pp. 15-21, 2008.

[22] C.-T. Huang, M. Matthews, M. Ginley, X. Zheng, C. Chen, and J. M. Chang, "Efficient and Secure Multicast in WirelessMAN:A Cross-layer Design," Journal of Communications Software and Systems, vol. 3, pp. 199-206, 2007.

[23] S. Chakraborty, S. Majumder, F. A. Barbhuiya, and S. Nandi, "A Scalable Rekeying Scheme for Secure Multicast in IEEE 802.16 Network," Communications in Computer and Information Science, Springer, vol. 132, pp. 471-481, 2011.

[24] J. Brown, X. Du, and M. Guizani, "Efficient rekeying algorithms for WiMAX networks," Security and Communication Networks, vol. 2, pp. 392–400, 2009.

[25] M. M. G. Sadeghi, B. M. Ali, M. Ma, J. A. Manan, N. K. Noordin, and S. Khatun, "Scalable Rekeying Algorithm in IEEE 802.16e," in 17th Asia-Pacific Conference on Communications (APCC), Sabah, Malaysia, 2011, pp. 726-730.

[26] J. A. Store, An Introduction to Data Structures and Algorithms. Waltham, USA: Birkhauser, Springer, 2001.